Two Methods FOR THE Exact Solution OF Diffraction Problems

Two Methods FOR THE Exact Solution OF Diffraction Problems

Frederick E. Alzofon

SPIE PRESS

A Publication of SPIE—The International Society for Optical Engineering
Bellingham, Washington USA

Library of Congress Cataloging-in-Publication Data

Alzofon, Frederick E.
    Two methods for the exact solution of diffraction problems / Frederick E. Alzofon.
        p. cm. – (SPIE Press monograph ; PM127)
    Includes bibliographical references and index.
    ISBN 0-8194-5141-X
        1. Electromagnetic waves—Diffraction. 2. Sommerfeld polynomial method. I. Title. II.
Series.

QC665.D5A49 2003
535'.42—dc21   2003057263

Published by

SPIE—The International Society for Optical Engineering
P.O. Box 10
Bellingham, Washington 98227-0010 USA
Phone: (1) 360.676.3290
Fax: (1) 360.647.1445
Email: spie@spie.org
Web: www.spie.org

Dedicated to

Professor Griffith C. Evans

Contents

Preface

This monograph is about two methods of calculating the electromagnetic fields due to radiation scattering by a single scatterer. Both methods yield valid results for all wavelengths of the incident radiation as well as a wide variety of scatterer configurations. Only the theory is discussed; numerical consequences of the theory are not presented.

Ruling out any changes of state owing to high-energy processes, two essential features of radiation scattering are: (1) the shape of the scatterer, and (2) the wavelength of the incident radiation. Taking these accurately into account in scattering analyses is a goal that has challenged researchers since the formulation of optical science. Many approximate methods have been devised to this end.

One of the methods to be generalized and discussed in this book was originated by Arnold Sommerfeld in two fundamental papers published in 1896 and 1897. The promising simplicity and accuracy of the expressions derived, well-verified by experiment, led to expectations of further progress in the use of Sommerfeld's method that were not realized in the years that followed. However, the need for exact solutions has been demonstrated by frequent references to the implications of Sommerfeld's work, even when the scatterers did not have Sommerfeld's original configuration. The lack of success in generalizing Sommerfeld's analysis has led to a general belief that it cannot be done.

The author was fortunate to find an alternate formulation of Sommerfeld's method—a logical continuation of research originated under the direction of Professor Griffith C. Evans at the University of California, Berkeley. This research was directed toward a generalization of Sommerfeld's method of constructing multiple-valued potentials defined on a three-dimensional multileaved space, i.e., solutions to Laplace's equation. Eventually, such a generalization was achieved and published in 1970. It was then evident that a similar generalization could be applied to construct multiple-valued Green's functions for the equation of wave propagation. The results constitute part of the work in the following text, indicating a variety of ways in which Green's functions can be constructed for a variety of configurations. For this introductory work, the method and the numerous consequences that follow from it render the theory more significant than the possible resulting calculations. However, it is hoped that such calculations will soon be carried out.

Although Sommerfeld's method delivers solutions in a convenient, closed, and analytic form, these solutions have the drawback of being applicable only to surfaces spanning given space curves; it is not applicable to solids bounded by closed surfaces. This limitation is remedied in a method of analysis originated by the author and discussed following the exposition of Sommerfeld's method. The new method provides a means of solving boundary-value problems for solids bounded by a large variety of surface configurations.

Although the analyses presented are primarily mathematical, they are not strictly rigorous. In this respect, the discussion follows the example set in the historical development of optics theory in which physical models and sufficient, rather than necessary, conditions have played an important role. The use of sufficient conditions, for example, is illustrated by superpositions of elementary solutions of the wave equation based on a physical model; it is then assumed that the series will converge to the correct solution of the boundary-value problem. Further, if the physical model is defined well enough to serve as a guide to any approximation, even a divergent series can be used as a solution, e.g., an asymptotic expansion. Indeed, Sommerfeld's original solution for diffraction by a semi-infinite plane was given as the first term of an asymptotic expansion. A similar expansion is used in the second method of this book, although it is not an essential part of the analysis.

It will be seen that the following work can be considerably extended.

Frederick E. Alzofon
September 2003

Chapter 1
Introduction

This monograph is concerned with two aspects of optical engineering: (1) the analysis of radiation diffraction valid for all wavelengths of the incident radiation, and (2) the analysis of radiation scattering by objects of arbitrary shape. It is a report of work in progress, oriented to the needs of optical engineers and its application in modern optical system design.

Since the analysis is valid for all wavelengths of incident radiation, it is worthwhile to distinguish between scattering and the diffraction of radiation. The term *scattering* generally refers to the interaction of a physical system with incident radiation, which results in an additional or scattered radiation field, while *diffraction* is a scattering process related to departures from geometrical optics[1] (as distinguished from physical optics). The two terms are used interchangeably throughout this text.

Separating geometrical and physical optics may not be acceptable to some optical engineers. Although geometrical optics is characterized by one text as that branch of optics for which the radiation wavelength is negligible,[2] a more recent opinion holds that it is a conceptual error to separate geometrical and physical optics.[3] In all cases, the wave nature of radiation should be of paramount concern; i.e., wave surfaces or their normals (rays) are of primary interest to the optical engineer.[3] This is the emphasis in the analyses to follow.

The need for more accurate and more easily calculated methods of analyzing diffraction problems has increased in recent years owing to the development of technologies that require a more reliable theoretical foundation. Among these are improved lens-design methods,[4] diffractive optics (holography),[5,6] compact-disk design,[7] optical computing,[8] millimeter and submillimeter technologies, and improved radar return predictions. Moreover, the establishment of close analogies between the solutions of wave propagation for nondispersive mediums and the solutions of parabolic differential equations for dispersive mediums (e.g., for diffusion and heat transfer in solids, as well as the classic Schrödinger equation) has shown the value of invoking optical concepts in various technologies, including thermography, hydrodynamics,[9–10] etc., as well as in radiation diffraction effects. As a result, the methods to be developed will have a wider application than for the electromagnetic field. It is hoped that the theory presented will aid in "raising the level of optical design."[3]

1.1 Sommerfeld's Method

Sommerfeld's method of solving diffraction problems by use of multiple-valued solutions of the wave propagation equation[11] is an extension of the electrostatic image method attributed to Lord Kelvin (also known as Sir William Thomson).[12] The results of Sommerfeld's analysis are delivered in closed form and are well suited to numerical calculation; the predictions of the method have been verified repeatedly by experiment.[13]

The agreement with experiment as well as the simplicity of the form of the solution has sometimes led to Sommerfeld's original expressions being used for configurations other than the one originally employed. In addition, the original Sommerfeld solution has been used to explain diffraction effects in even more general contexts: for example, to decide in favor of Thomas Young's proposal of a reflected boundary wave.[2]

Despite its promising beginning, the Sommerfeld method of analyzing diffraction problems is not recognized as being capable of generalization. Although the results of the original investigation are frequently quoted, there is an accepted opinion that the method cannot be generalized in any useful manner.[14] However, much of this book demonstrates that such a generalization is possible.

Certainly, there are already methods of predicting the effects of diffraction and scattering. For example, the Kirchhoff integral for analyzing the diffraction of scalar and vector fields is well known and frequently used.[1,2] However, it is admitted in Ref. [1] that the mathematical assumptions at the basis of the latter applications are flawed and do not agree with the physical models, even when wavelengths are small compared with the diameter of the diffracting obstacle or aperture. For example, a comparison of the scalar theory with observation shows that theory is very far from agreement with experiment.[4] Rigorous treatments of diffraction, such as the dual integral equations method, are very difficult to implement.[2] The drawbacks of other methods suggest that it is worthwhile to develop a promising alternative approach, i.e., the Sommerfeld method.

However, as already noted, the original Sommerfeld method of solving a diffraction problem is not easy to generalize. Consequently, an alternative method—devised by Sommerfeld[12] and used to construct multiple-valued solutions of Laplace's equation—can construct solutions of diffraction problems. This approach is suggested by the successful generalization of the methods discussed in Refs. [12] and [15].

Thus, one begins with the Green's function for a single-leaved space, which is a solution of the wave propagation equation, and proceeds in a manner similar to Ref. [15]. The result is a spherical incident wave rather than a plane wave; therefore, there is Fresnel diffraction rather than Fraunhofer diffraction, as in Ref. [12]. However, it will be shown that a multiple-valued spherical wave with a source point at a great distance from the observer is essentially a plane wave. It follows that expressions can be derived for multiple-valued plane waves from those for multiple-valued spherical waves. Thus, Fraunhofer diffraction can be analyzed from the formalism for Fresnel diffraction.

After a review of the historical background of Sommerfeld's method in Chapter 2, Chapter 3 demonstrates the construction of a multiple-valued Green's function for a two-leaved space bounded by one branch line. In Chapter 4, this function is applied to Fresnel diffraction by a perfectly conducting half-plane. Chapters 5 through 7 apply the same procedure to Fresnel diffraction by a perfectly conducting circular disk, a circular annulus, and a slit between two half-planes.

The scattering configurations listed above can be generalized by altering the coordinate systems used to define the original spaces. This aspect of Sommerfeld's method is discussed in Chapter 8 along with an alternate means of constructing new coordinate systems to be used in Chapters 9 and 10.

1.2 Generalizing Boundary Surfaces

A considerable drawback to solving radiation scattering problems has been the lack of analytical tools to deal with solids bounded by nonequipotential surfaces. This circumstance has led to the use of equivalent Mie scatterers for nonspherical solids, for lack of a better alternative. In this respect, the solution of a boundary-value problem for radiation scattering by a sphere, i.e., Mie scattering, plays a role similar to Sommerfeld's original solution for diffraction by a half-plane, since it also has been applied to configurations other than the one considered in the original analysis.

In response to the need for alternate methods of solving boundary-value problems for radiation scattering, many kinds of solutions have been developed; these have depended on the availability of computers (Appendix A). Numerical methods have often been tailored for specific kinds of problems.

Although the Sommerfeld method can provide unique and valuable analyses of radiation scattering, it has several drawbacks, one of which is its applicability solely to surfaces that are not closed (i.e., without an interior and an exterior). The results are not restricted with respect to the relative magnitudes of wavelength and scatterer diameter, but there is an implicit restriction in any experiment based on the method with respect to the relative magnitudes of wavelength and thickness of the scatterer; i.e., their ratios must be very small.

Sommerfeld's method, which imagines scatterers with "knife edges," leads to the possibility of radiation sources on the branch curves bounding the scattering surfaces. In this event, there may be a violation of the condition for uniqueness of any solution derived.[2,16] Although it will be shown that this circumstance does not occur with the expressions to be derived, it is advisable to develop an alternate method of solving boundary-value problems without such restrictions.

Chapters 9 and 10 offer this type of alternate method of solution of boundary-value problems, based on the introduction of coordinate systems that are dependent on transformations of similitude[17] of the given boundary surfaces. As a concrete example of this method, the details of constructing a solution for the case of a plane wave scattered by a hexagonal ice cylinder are presented.

Four appendixes contain a list of other exact methods for solving diffraction problems (Appendix A), a brief review of Sommerfeld's original work (Appendix B), and some mathematical theory (Appendixes C and D) as an aid in the analyses of the earlier chapters.

References

1. J. Jackson, *Classical Electrodynamics, Second Edition*, p. 427, John Wiley & Sons, New York (1975).
2. M. Born and E. Wolf, *Principles of Optics, Sixth Edition*, Pergamon Press, New York (1980).

3. D. Sinclair, "Whither optical design?," *Optics and Photonics News*, **11**, pp. 34–35, 38–39 (2000).

4. G. Slyusarev, *Aberration in Optical Design Theory*, Adam Hilger, Ltd., Bristol, UK (1984).

5. J. DeVelis and G. Reynolds, *Theory and Applications of Holography*, Addison-Wesley, Menlo Park, CA (1967).

6. R. Collier, C. Burckhardt, and L. Lin, *Optical Holography*, Academic Press, New York (1971).

7. C. Bouwhuis, J. Braat, A. Huijser, J. Pasman, G. van Rosmalen, and K. Schouhamer Immink, *Principles of Optical Disc Systems*, Adam Hilger, Ltd., Bristol, UK (1985).

8. N. Kopeika, *A System Engineering Approach to Imaging*, SPIE Press, Bellingham, WA (1998).

9. F. Alzofon, "Optical analogues in a dispersive medium," Parts I and II, *J. Wave-Mat. Int.*, **8**, pp. 185–217 (1993).

10. F. Alzofon, "Some optical concepts in the analysis of viscous fluid flow," *J. Wave-Mat. Int.*, **11**, pp. 219–233 (1996).

11. A. Sommerfeld, "Mathematische Theorie der Diffraction," *Math. Ann.*, **47**, pp. 317–374 (1896).

12. A. Sommerfeld, Über verzweigte Potentiale im Raum," *London Math. Soc. Proc.*, **28**, pp. 395–429 (1897).

13. H. Hönl, A. Maue, and K. Westphal, "Theorie der Beugung," *Encycl. Phys.*, *Crystal Optics Diffraction*, S. Flügge, Ed., **25**(1), Springer Verlag, Berlin (1966).

14. D. Jones, *The Theory of Electromagnetism*, p. 574, Pergamon Press, New York (1964).

15. F. Alzofon, *Multiple-Valued Functions in Three-Dimensional Space and Sommerfeld's Method*, Lockheed Electronics Company, Houston, TX (1970).

16. A. Sommerfeld, *Partial Differential Equations in Physics*, pp. 188–193, Academic Press, New York (1949).

17. W. Osgood and W. Graustein, *Plane and Solid Analytic Geometry*, The Macmillan Company, New York (1938).

Chapter 2
Historical Background of the Sommerfeld Method

The Sommerfeld method of solving diffraction problems[1] is essentially a way of extending the range of applications of the image method originated by Lord Kelvin.[2] For this reason, some insight into the Sommerfeld method can be gained through a brief review of the Kelvin image method.

2.1 The Kelvin Image Method

Kelvin's image method extends the region on which a solution to the given problem is originally defined in order to make it easier to represent the solution in part of the original region. However, the new solution may not conform to physical reality in all of the extended region. To illustrate the essentials of this approach, a simple example can be presented that emphasizes those aspects relevant to the Sommerfeld method.

Consider an infinite, perfectly conducting grounded plane conductor of electrical charge, which coincides with the (x, z) plane of a rectangular Cartesian coordinate system $\{(x, y, z)\}$, as illustrated in Fig. 2.1. A point electric charge $+q(\text{esu})$ lies on the point $(0, +d, 0)$. The goal of this analysis is to find the potential function of for the entire space. Since the region specified by $y < 0$ is shielded from the electric charge $+q$, the electric field must be zero in this region, and therefore the potential function is a constant—for example, equal to zero for $y < 0$. Since no charge is in motion on the conducting plane, the electric field is zero on the plane and the potential function can also be chosen as equal to zero for $y = 0$. Finding the potential function in the region $y > 0$ remains; this is done using the image method.

To this end, an artifice is employed. The original configuration is replaced by two point charges: the original charge $+q$ at $(0, +d, 0)$, and a charge $-q$ at $(0, -d,$

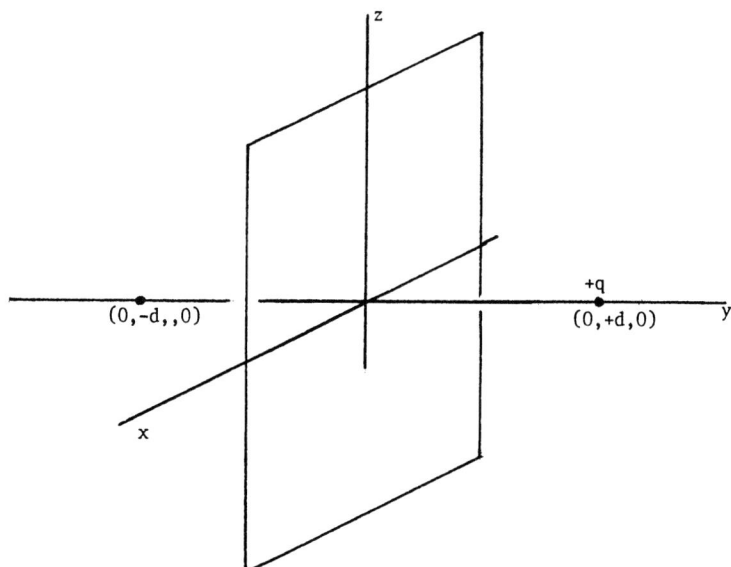

Figure 2.1 A simple case solvable by the Kelvin image method: a grounded plane conductor with one point charge, +q, placed before it.

0) after removing entirely the conducting plane at $y = 0$. The charge $-q$ is called the "image" of the charge $+q$, and the half space $y < 0$ where it lies is called the *image space*. In electrostatic units (d is measured in cm), the potential function corresponding to the imagined configuration at the point of observation (x, y, z) is

$$V = \left(\frac{q}{r_1}\right) - \left(\frac{q}{r_2}\right), \tag{2.1}$$

where $r_1^2 = x^2 + (y-d)^2 + z^2$ and $r_2^2 = x^2 + (y+d)^2 + z^2$. It is clear that, leaving aside the question of whether V is a solution of an appropriate differential equation, the function V vanishes, as it should, on the plane $y = 0$. Moreover, although V represents the original potential in the half space $y \geq 0$, it does not represent the potential for the original configuration in the image space $y < 0$.

In general, the Kelvin image method is limited to surfaces that divide all space into mutually exclusive, nonoverlapping regions that fill all of space: an infinite plane, an infinite wedge made up of two intersecting half-planes, a rectangular box, a sphere, an ellipsoid, etc. Sommerfeld modified Kelvin's method to avoid this limitation. Contrary to Kelvin's treatment, as shown in Chapter 4, an image source point is not introduced, although the imaging concept is retained.

2.2 The Sommerfeld Image Method

In the Sommerfeld image method, the surface (finite or semi-infinite in extent and spanning a given space curve) on which boundary conditions are to be satisfied is imagined to divide all of ordinary or physical space (of infinite extent) from an imagined three-dimensional image space (also of infinite extent). To clarify the concept, the ordinary space—specified by, for example, a cylindrical coordinate system $\{(r, \phi, z)\}$, where $0 < r < +\infty$, $0 \leq \phi < 2\pi$, $-\infty < z < +\infty$—is joined to an image space specified by $0 < r < +\infty$, $2\pi \leq \phi < 4\pi$, $-\infty < z < +\infty$. The usefulness of the augmented space arises from the property that a function varying as $\exp(i\phi / 2)$ becomes single-valued on the space; in such a case, the half-planes $\phi = 0$ and $\phi = 4\pi$ are considered to be the same half-planes—they are "identified." The sum or union of the two regions is called a *two-leaved space*; the regions or leaves are separated by the half-plane $\phi = 0$ (or by $\phi = 4\pi$). The z-axis is called a *branch line* and bounds both leaves but does not belong to either leaf.

It is not necessary that the branch curve bounding both leaves be infinite in extent or that the surface separating the leaves be infinite or even semi-infinite. For example, in Fig. 2.2 a two-leaved space is depicted. The leaves are bounded by a circle centered on the z-axis and lying in the (x,y) plane. The surface separating the leaves may be chosen in a variety of ways; in all cases, the surface must span the branch curve. The separating surface can be a flat, circular disk as in Fig. 2.2; a spherical cap; or the section of an infinite plane extending from the boundary circle to infinity. When one of these surfaces is chosen by the analyst to be the one on which boundary conditions are imposed, it constitutes an obstacle to the incident radiation. The disk belonging to the first leaf in Fig. 2.2 can be imagined to be such an obstacle, or, in a useful duality, it can be imagined to be an aperture in the rest of the plane surrounding the aperture, depending on the boundary conditions.

A continuous circuit described by a point in the two-leaved space, linking the branch circle, takes the point from the first leaf through the points numbered in sequence into the second leaf and then again into the first leaf. Circuits in each leaf not linking to the branch curve leave the variable point within the leaf, also shown in Fig. 2.2.

With the aid of such concepts, Sommerfeld solved the wave equation, describing the diffraction of a plane wave by a perfectly conducting half-plane (Fraunhofer diffraction) valid for all values of the wavelength of the incident radiation.[1] Yet, despite the promise of this study, texts on the electromagnetic theory imply, nearly a century later, that Sommerfeld's method has no useful generalization.[3,10] However, even without a useful generalization, in the years since the original work Sommerfeld's results for the half-plane have

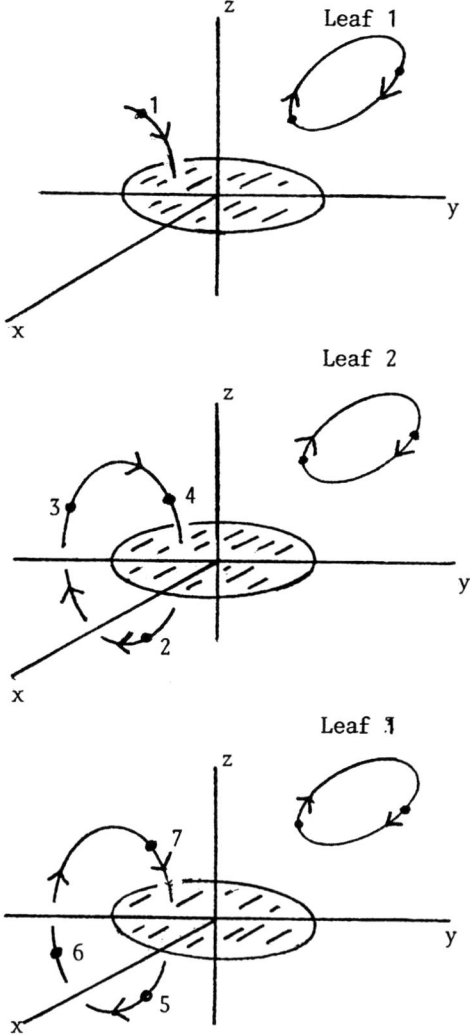

Figure 2.2 The result of continuous paths linking and not linking a branch curve in a two-leaved space.

been cited in connection with different diffracting configurations—simply because the analysis was trustworthy for all wavelengths of the incident radiation.

Thus, under more general conditions, the Sommerfeld solution is used to justify the assumption that diffraction can be accounted for as the combined effect of an incident wave and a boundary wave.[3] This assumption has led to fruitful investigations.[4,5] In addition, Sommerfeld's solution is sometimes applied in a semiquantitative manner to configurations other than the semi-infinite plane, as a reliable standard of comparison.

Despite the acknowledged value of the Sommerfeld solution, no systematic approach to a solution by the same method for configurations other than a half-plane is well known. In addition, there is a belief that the Sommerfeld solution "must be regarded as a triumph of mathematical ingenuity and experiment."[6] A similar evaluation is to be found in Ref. [7], which views Sommerfeld's solution as the result of experiment rather than a systematic method. Consequently, Wiener-Hopf integral equations are sometimes used[8] as well as other approximate, computer-aided methods (see Appendix A).

However, the need for more satisfactory treatments of diffraction problems, such as in lens design, continues to grow as applications of optics expand. A recent paper by an optical designer asserts that both geometric and physical optics require overlapping theoretical treatments. For example, lenses may be placed in spaces that have dimensions of only a few wavelengths, requiring allowance for diffraction effects.[9]

Contrary to a text on the theory of electromagnetism,[10] it will be shown in the following chapters that the Sommerfeld method can be significantly generalized. Characteristically, the Sommerfeld solutions are delivered in closed form and readily lend themselves to evaluation in numerical terms. They are valid for all values of the wavelength of incident radiation and for a wide class of configurations. The discussion of these solutions is an expanded version of Refs. [11] and [12].

Although the emphasis in this book is on radiation scattering, it should be mentioned that the Sommerfeld method is useful in a variety of other applications of interest to the engineer and scientist. Applied to static fields, it represents a limiting case for which the wavelength of the incident radiation is infinite. The static-field solutions (of Laplace's equation) are therefore of interest as a check on the accuracy of derivations of nonstatic fields. Moreover, the static-field solutions are of interest in their own right. For example, with the aid of multiple-valued solutions of Laplace's equation, surfaces of minimum capacity and minimum induction can be derived and are useful in the study of the shapes assumed by rotating fluids in an equilibrium state.[13,14]

Using multiple-valued solutions, Sommerfeld derived the conductor potential for a semi-infinite plane bounded by one straight line.[2] Using a similar analysis, Hobson obtained the conductor potential for a circular disk, a study extended in Ref. [14]. Evans proved the existence of theorems for potential solutions defined on multileaved spaces in Ref. [15], ensuring the existence of solutions under general conditions.

There is little doubt that further uses of multiple-valued solutions of differential equations will be found.

References

1. A. Sommerfeld, "Mathematische Theorie der Diffraction," *Math. Ann.*, **47**, pp. 317–374 (1896).

2. A. Sommerfeld, "Über verzweigte Potentiale im Raum," *London Math. Soc. Proc.*, **28,** pp. 395–429 (1897).

3. M. Born and E. Wolf, *Principles of Optics, Seventh (expanded) Edition*, pp. 499–500, Cambridge Univ. Press, Cambridge, UK (1999).

4. A. Rabinowicz, *Die Beugungswelle in der Kirchhoffsche Theorie der Beugung, Second Edition*, Springer Verlag, Berlin (1966).

5. A. Rabinowicz, *Progress in Optics*, E. Wolf, Ed., **4**, John Wiley & Sons, New York (1965).

6. H. Bateman, *Partial Differential Equations of Mathematical Physics*, p. 477, Dover Publications, New York (1944).

7. *Handbuch der Physik*, Band XXV/1, p. 222, S. Flügge, Ed., Springer Verlag, Berlin (1961).

8. B. B. Baker and E. T. Copson, *The Mathematical Theory of Huygens Principle, Second Edition*, Oxford Univ. Press, London (1950).

9. D. C. Sinclair, "Whither optical design?," *Optics and Photonics News*, **11**, pp. 34–39 (2000).

10. D. S. Jones, *The Theory of Electromagnetism*, p. 574, Pergamon Press, New York (1964).

11. F. Alzofon, "Multiple-valued solutions of the equations of field propagation: Sommerfeld's method, I–II," *Wave-Mat. Int.*, 7, pp. 121–135 (1992).

12. F. Alzofon, "Multiple-valued solutions to the equations of field propagation: Sommerfeld's method, III, Fraunhofer diffraction," pp. 1–14, and "Multiple-valued solutions to the equations of field propagation: Sommerfeld's method, IV, Fresnel diffraction," pp. 89–104, *J. Wave-Mat. Int.*, 8 (1993).

13. H. Poincaré, *Figures d'Équilibre d'une Masse Fluide*, Gauthier-Villars, Paris (1902).

14. E. W. Hobson, "On Green's function for a circular disc, with applications to electrostatic problems," *Cambridge Philosophical Trans.*, 18, pp. 277–291 (1900).

15. G. C. Evans, "Lectures on multiple-valued harmonic functions in space," *University of California Publications in Mathematics*, University of California Press, New Series, 1, pp. 281–340 (1951).

16. S. F. Neustadter, "Multiple-valued harmonic functions with circle as branch curve," *University of California Publications in Mathematics*, University of California Press, New Series, 1, pp. 396–432 (1951).

Chapter 3
Two-Leaved Generalization of a
Spherical Wave: One Branch Line

In this chapter, the Sommerfeld method is applied to the construction of a solution of the wave propagation equation defined on a two-leaved space bounded by a straight line. The solution is a two-leaved generalization of a spherical wave, analogous to Sommerfeld's generalization of a plane wave in physical space. (For the sake of comparison with Sommerfeld's original work, a very brief summary of his original paper is given in Appendix B.) It is then possible to solve a boundary-value problem for a spherical electromagnetic wave incident on a perfectly conducting semi-infinite plane, which will be done in Chapter 4.

A spherical wave solution (Fresnel diffraction) lends itself to experimentation more readily than Sommerfeld's solution (Fraunhofer diffraction) since the generation of a spherical wave in the laboratory does not require a collimating lens to produce the incident radiation, providing a considerable advantage.[1] Another advantage of the spherical wave generalization lies in that it is generated by a point source, and like its analog in the theory of static electricity, it can be used to construct extended source distributions. Finally, if the source point of a spherical wave is imagined to be very far from the point of observation, a plane wave is approximated and the corresponding boundary-value problem represents Fraunhofer diffraction. These fundamental ideas are developed in this chapter and applied to subsequent applications of Sommerfeld's method.

3.1 The Point Radiation Source in Physical Space

The scalar function often used to represent a point radiation source in ordinary (physical) space is $[\exp(ikD)]/D$, where $D = D(P, P')$ is equal to the distance between a fixed source point P' and the point of observation P. The latter source function is a solution of the equation of wave propagation in a vacuum (sometimes called the Helmholtz equation):

$$\left(\Delta + k^2\right)U = 0,\tag{3.1}$$

where Δ is the Laplacian operator and $k = 2\pi/\lambda$, where λ is equal to the wavelength of the radiation in a vacuum.

Sommerfeld's method, when applied to static fields, i.e., to Green's functions that are solutions of Laplace's equation

$$\Delta V = 0,\tag{3.2}$$

e.g., $V = 1/D$, lead to potentials defined on multileaved spaces.[2,3] In contrast, the Green's function $U = \exp(ikD)/D$ is composed of the sum $[(\cos kD)/D] + i[(\sin kD)/D]$, where only the first term can be called a source function, i.e., varying as $1/D$ for values of D close to zero. Thus, for the purpose of analysis, it may at times be more relevant to the goal of extending the Sommerfeld method to use the function $(\cos kD)/D$ rather than $\exp(ikD)/D$.

In any case, note that the distance function D plays an essential role in Sommerfeld's method. For both physical and multileaved spaces it has the same form; in terms of a Cartesian rectangular-coordinate system $\{(x, y, z)\}$, it can be written that

$$D = +\sqrt{\left(x - x'\right)^2 + \left(y - y'\right)^2 + \left(z - z'\right)^2},\tag{3.3}$$

and in terms of a cylindrical coordinate system $\{(r, \phi, z)\}$, it can be written that

$$D = +\sqrt{\left(r - r'\right)^2 + \left(z - z'\right)^2 + 2rr'\left[1 - \cos\left(\phi - \phi'\right)\right]}.\tag{3.4}$$

3.2 Complex Number Notation

The analysis to follow is considerably simplified if complex numbers and the theory of analytic functions are used. Some of the theory of analytic functions of a complex variable is presented in Appendix C of this volume, and references are cited there for further study of the theory.

In addition to the (x,y)-plane, the set of complex numbers $\rho = x + iy$ is considered, where $i = \sqrt{-1}$. The set of complex numbers is called the ρ-plane. In terms of this notation, Eq. (3.3) can be written in the abbreviated form

$$D = +\sqrt{\left|\rho - \rho'\right|^2 + \left(z - z'\right)^2} \ . \tag{3.5}$$

Moreover, for a cylindrical coordinate system, setting $\rho = r + iz$, Eq. (3.4) can be written in the form

$$D = +\sqrt{\left|\rho - \rho'\right|^2 + 2\Re\rho\Re\rho'\left[1 - \cos\left(\phi - \phi'\right)\right]}, \tag{3.6}$$

where $\Re\rho = r$, the "real part" of $\rho = r + iz$. In addition, note that if $\rho = x + iy$, then $r = +\sqrt{x^2 + y^2} = \left|\rho\right|$ and $\rho = r\exp i\phi = r\cos\phi + ir\sin\phi$.

For reasons that will become clearer in subsequent chapters, it is convenient to introduce $\{(\eta, \phi, z)\}$ as a new set of coordinates, where

$$r = a\exp\left(-\eta\right), \ \left(-\infty < \eta < +\infty\right), \tag{3.7}$$

and a is a positive constant (Fig. 3.1). The surfaces η = constant are circular cylinders: $\eta = 0$ is the cylinder $r = a$, while $\eta < 0$ corresponds to the exterior of the latter cylinder; $\eta > 0$ represents the interior of that cylinder (Fig. 3.2). It also follows that $\rho = a\exp(-\eta + i\phi)$, and

$$D = +a\sqrt{2}\left[\exp\frac{\left(-\eta - \eta'\right)}{2}\right]\sqrt{\cosh\alpha - \cos\left(\phi - \phi'\right)}, \tag{3.8}$$

where $\cosh\alpha = \cosh(\eta - \eta') + (z - z')^2 / 2rr'$, and $rr' \neq 0$. The quantity α can vary from 0 to $+\infty$; the z-axis has been removed from the space.

As defined earlier, the physical or ordinary space (leaf 1, or $0 \leq \phi < 2\pi$) is joined to a second leaf, specified by the interval $2\pi \leq \phi < 4\pi$. The union of the two leaves $\{(\eta, \phi, z)\}$, $0 \leq \phi < 4\pi$ forms a two-leaved space bounded by the branch curve $r = 0$. The latter branch curve bounds both leaves. It is noted that the underlying points (η, ϕ, z) and $(\eta, \phi, + 2\pi, z)$ are separated by the distance $D = 0$, although they are distinct points and lie in different leaves.

3.3 Outline of the Construction of a Multiple-Valued Radiation Source

This section lists a summary of the method of constructing a multiple-valued, radiation-source solution so that the orientation of the following discussion can be preserved amid detailed calculations.

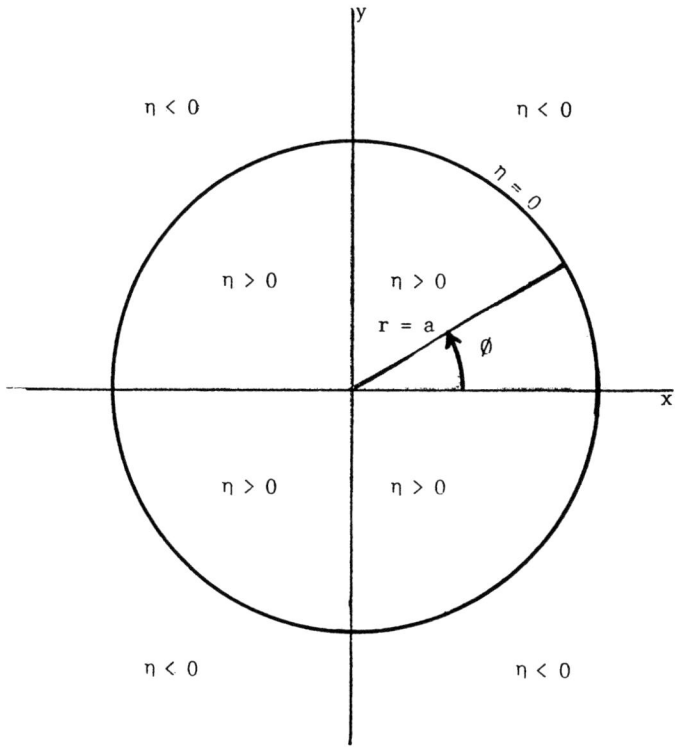

Figure 3.1 Regions of the (*x,y*)-plane specified by the algebraic sign of η .

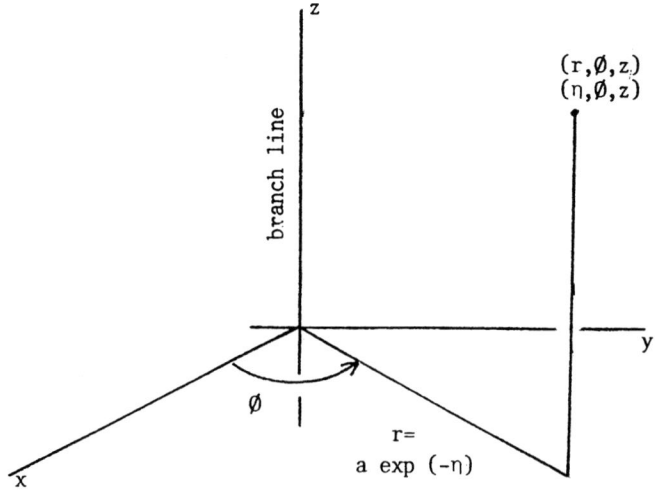

Figure 3.2 Relation between the cylindrical and the $\left\{\left(\eta,\phi,z\right)\right\}$ coordinate system.

(1) The dependence of $(\exp ikD)/D$ on the fixed angle ϕ' is altered by an analytic continuation (see Appendix C) of ϕ' to the variable $u = \phi'' + i\eta''$; the latter variable will eventually serve as a variable of integration in a Cauchy integral. Upon applying the same process of analytic continuation to both sides of the equation of wave propagation, Eq. (3.1) demonstrates that the resultant function remains a solution to the wave equation as a function of the variable point P.

(2) The function resulting from the continuation in (1) is denoted by U, and can now be written as a Cauchy integral over a contour in the u-plane with an explicit dependence on ϕ' through the term $\exp(i\phi'/2)$.

(3) The analytic continuation of ϕ' has caused D to become a complex function with branch points in the u-plane. In such a circumstance, the contour of the Cauchy integral cannot be deformed at will unless barriers are introduced in the u-plane to prevent a circuit of a branch point, preserving the single-valued character of the integrand in the Cauchy integral.

(4) Once having introduced the barriers, the integrand of the Cauchy integral can be exhibited as an explicit function of $\phi/2$; the integral is still, however, a single-valued function of ϕ, with period 2π.

(5) Part of the contour of integration is now deleted and a new function U_1 is defined.

(6) It is shown that the new function $U_1(P, P')$ is a solution of the wave equation as a function of P and is single-valued in a two-leaved space bounded by the branch curve $r = 0$ and the point P' (which has been removed from the space by the condition $D \neq 0$). In addition, U_1 behaves essentially as $(1/D)\cos kD$ or $(1/D)\exp ikD$ in the neighborhood of P'. The function U_1 approaches zero as D increases without bound for P in either leaf of the two-leaved space. Because of the latter properties, it may be said that U_1 is a two-leaved generalization of the point radiation source $(\cos kD)/D$ or $(\exp ikD)/D$.

3.4 The Analytic Continuation of ϕ'

To accomplish the analytic continuation of ϕ', it is replaced by $u = \phi'' + i\eta''$, where ϕ'' and η'' are solutions of the Cauchy-Riemann equations (discussed in Appendix C); they are conjugate functions. Because of this transformation, $\rho' = a \exp i(\phi' + i\eta')$ is altered to $\sigma = a \exp i(\phi'' + i\eta'' + i\eta')$; the complex conjugate of $\rho', \rho'* = a \exp -i(\phi' - i\eta')$ is altered to $\delta = a \exp(-\eta' + \eta'' - i\phi'')$. Figures 3.3(a) and (b) illustrate these transformations. The absolute values of σ and δ are related by an inversion in a circle of radius $a \exp(-\eta')$, i.e., $|\sigma||\delta| = a^2 \exp(-2\eta')$.

When the above transformation is performed on the distance function D, it is altered to D_1, where

$$D_1 = a\sqrt{2} \left[\exp \frac{(-\eta - \eta')}{2} \right] \sqrt{\cosh \alpha - \cos(u - \phi)} \,. \qquad (3.9)$$

The branch points of D_1, where $D_1 = 0$, are $u = u_b = \phi \pm i\alpha + 2\pi p$, where $p = 0, \pm 1, \pm 2, \pm 3, \ldots$ The branch of the square root of D_1 that reduces to D, which is a positive quantity if $u = \phi'$, is chosen.

Alternatively, setting $\delta = a^2[\exp(-2\eta')]/\sigma$, it can be written that

$$D_1^2 = \left(\frac{-\rho *}{\sigma} \right)(\sigma - \sigma_{b1})(\sigma - \sigma_{b2}), \qquad (3.10)$$

where

$$\left(\frac{\sigma_b}{\rho} \right) = \left(\frac{1}{2} \rho\rho * \right) \left\{ \rho'\rho' * + \rho\rho * \pm \sqrt{\left[\rho'\rho' * + \rho\rho * + (z - z')^2 \right] - 4|\rho\rho'|^2} \right\} \qquad (3.11)$$

and

$$\sigma_b = a \exp\left(-\eta' \mp \alpha + i\phi \right). \qquad (3.12)$$

It follows that a circuit of the z-axis by the point P generates a circuit of the origin of the ρ-plane (or z-axis) by both σ_{b1} and σ_{b2}.

The analytic continuation of ϕ' has resulted in the creation of an infinite number of branch points: $u_b = \phi \pm i\alpha + 2\pi p (p = \pm 1, \pm 2 \ldots)$ in the u-plane. A circuit of one of these points in the u-plane alters the algebraic sign of D_1, and as a result, D_1 is not single-valued in the u-plane. A continuous deformation of the

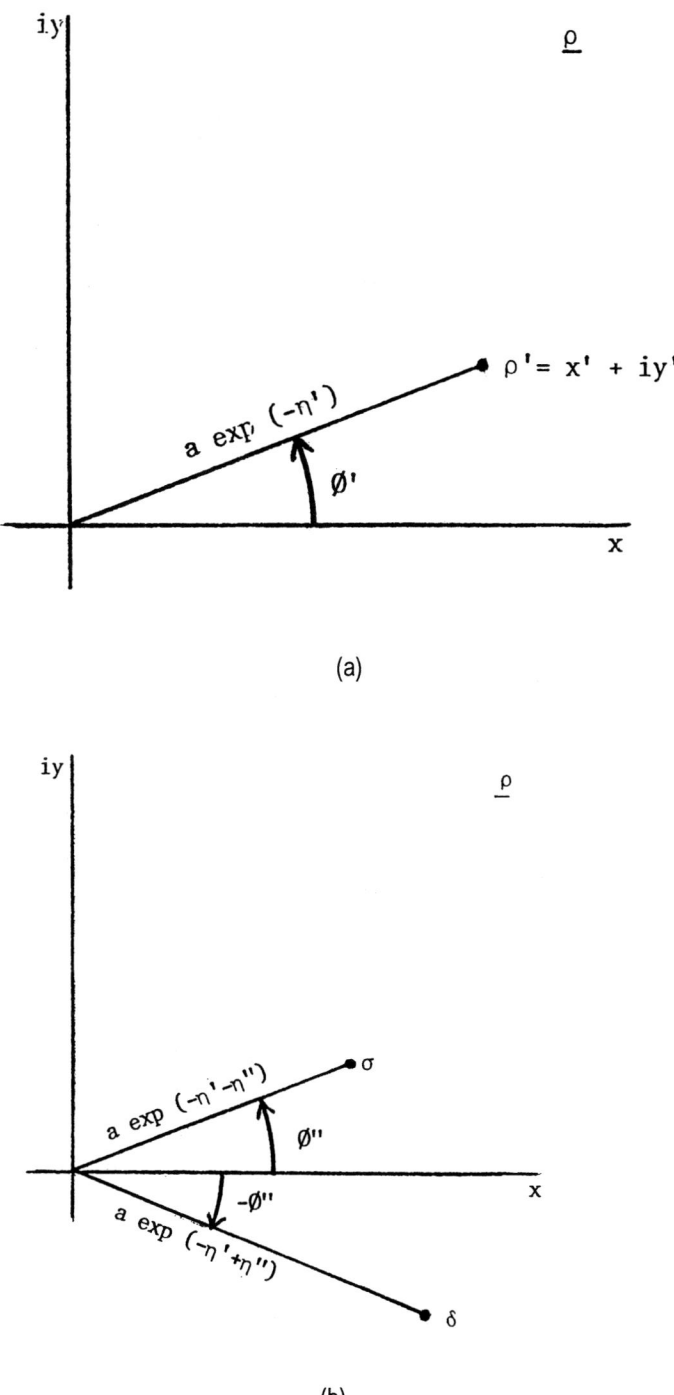

(a)

(b)

Figure 3.3 (a) The fixed point ρ'. (b) The transform of ρ', or σ, and the transform of $\rho' *$, or $\delta : \sigma\delta = a^2 \exp(-2\eta')$.

contour for the Cauchy integral to be defined would not be permissible because the contour would be deformed over one of the branch points. This difficulty can be avoided by joining the branch points in pairs by barriers through which the contours cannot be deformed (Fig. 3.4); the points of the barriers are removed from the u-plane. The barriers can join the branch points to each other or to the point at infinity (see Figs. 3.5 and 3.6) so that (1) the value of D_1 will be returned to its original value upon a circuit, or (2) so that no single branch point in the u-plane can be circuited.

Unfortunately, the first case cannot be used since, in the construction of the Cauchy integral yet to be defined, the point $u = \phi'$ must not be allowed to lie upon the branch cut barrier. Consequently, only the barriers illustrated in Figs. 3.5 and 3.6 will be used; they have been featured in Refs. [2]–[4]. Sommerfeld's original analysis will be followed as closely as possible.

3.5 The Cauchy Integral for a Point Source: Definition of U_1

An essential step in the Sommerfeld method is the representation of the radiation point source function as a Cauchy integral (see Appendix C). This enables the original dependence of the source solution on ϕ to be altered to a dependence on $\phi/2$, eventually creating a function defined on a two-leaved space.

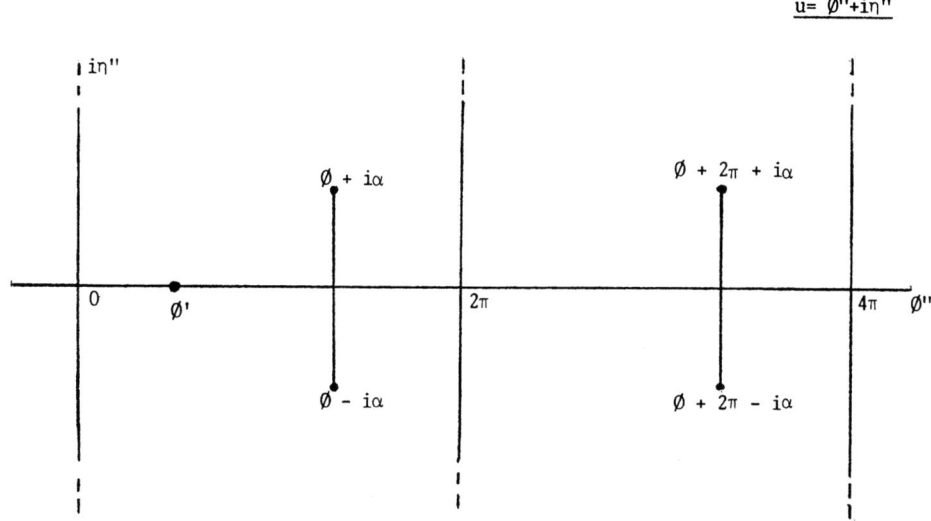

Figure 3.4 Barriers in the u-plane connecting the branch points in pairs $(\alpha \neq 0)$.

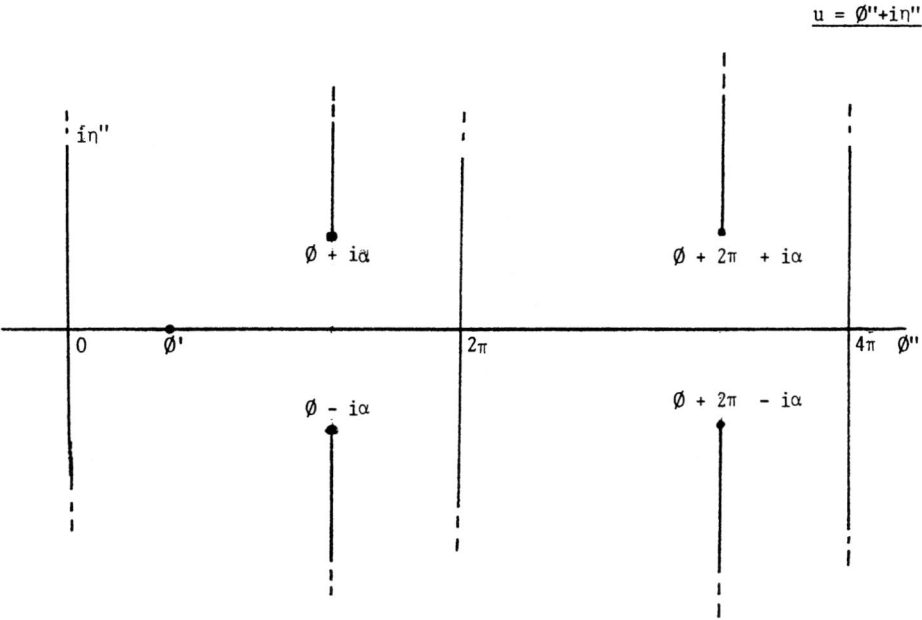

Figure 3.5 Barriers in the *u*-plane connecting each branch point to the point at infinity $(\alpha \neq 0)$.

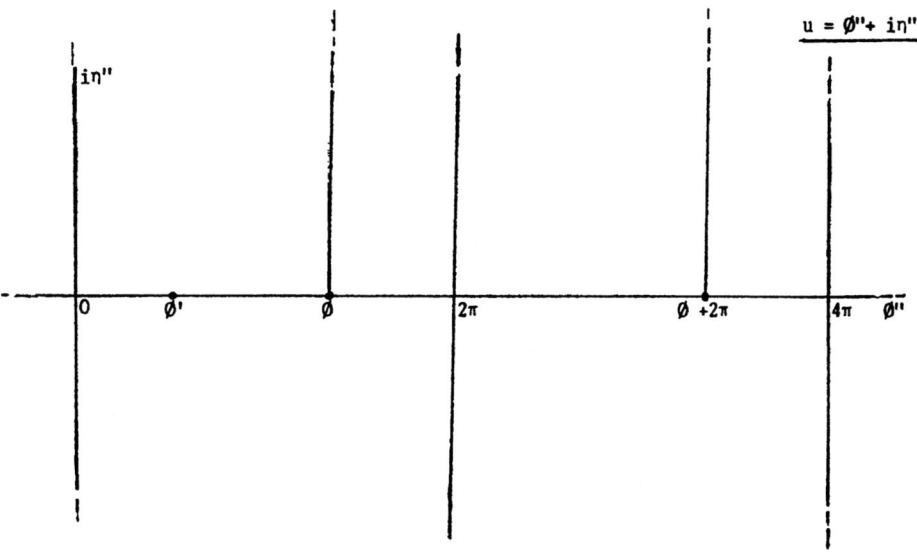

Figure 3.6 Barriers in the *u*-plane connecting a branch point to the point at infinity $(\alpha = 0)$.

It is easy to see that with P' in the first leaf (physical space: $0 \le \phi' < 2\pi$),

$$U = \frac{(\exp ikD_1)}{D} = \left(\frac{1}{4\pi}\right) \int_C \frac{du}{1 - \exp i(\phi' - u)/2} \cdot \frac{(\exp ikD_1)}{D_1}, \qquad (3.13)$$

where the contour of integration is a small circle in the u-plane with the point $u = \phi'$ in its interior (see Fig. 3.7). The interior of the circle must not contain any other singularities.

The contour is now deformed into the contour C_1 as shown in Fig. 3.8. Horizontal sections of the contour are described by $u = \phi'' + i\eta_1''$ and $u = \phi'' - i\eta_2''$, where both η_1'' and η_2'' are positive real numbers. Since the integrand is periodic in u, with period 4π, the contributions to the integral by the vertical sections of the contour lying on points $u = 0$ and $u = 4\pi$ cancel one another for all values of η_1'' and η_2''. Consequently, these sections of the contour can be deleted without altering the value of the integral.

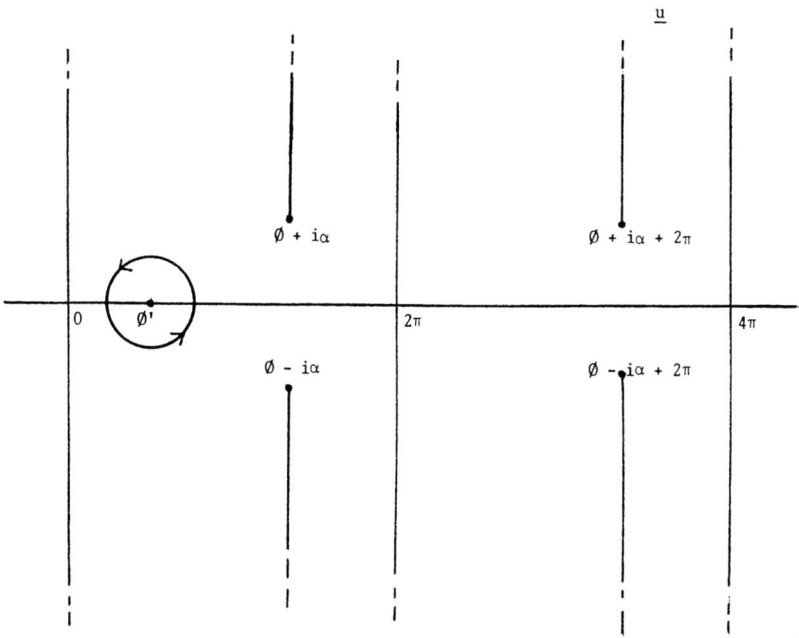

Figure 3.7 Small-circle contour for the Cauchy integral $(\alpha \neq 0)$.

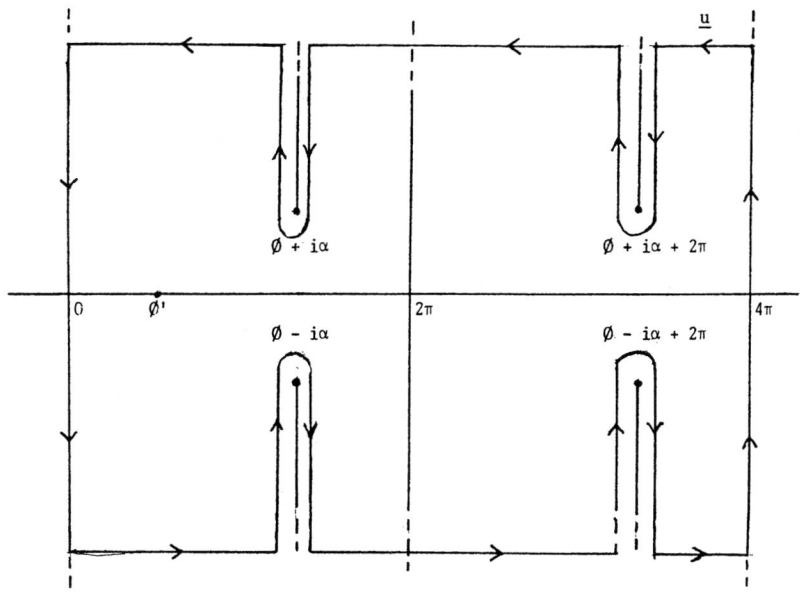

Figure 3.8 Contour C_1, equivalent to the small-circle contour in Fig. 3.7 $(\alpha \neq 0)$.

For large values of $|\eta''|$, D_1 behaves approximately as $a\sqrt{2}\,[\exp-(\eta+\eta')/2]\left(\pm i\sqrt{\cosh \eta''}\right)$. The change in sign occurs when the contour loops about a branch point. As a result, the term $(\exp ikD)$ may increase or decrease as the horizontal sections of the contour C_1 recede to infinity. For this reason, the integrand in Eq. (3.13) is altered by multiplying by factor g, which may be dependent on $u, \eta', \phi,$ and z' but not on the coordinates of P; g diminishes rapidly enough to cause the integrand to converge exponentially to zero (see Appendix D) as a function of η_1'' and η_2''. Such a factor must reduce to unity if $u = \phi'$ to preserve the value of the Cauchy integral. It must also reduce to the same value if $k = 0$ to preserve the physical meaning of the expression. Since g is a multiplicative factor, the integrand in Eq. (3.13) remains a solution of the wave propagation equation. It is readily seen that the choice of $g = \exp - ka[\cos(u - \phi') - 1]$ satisfies the conditions listed in Sec. 3.3.

The horizontal sections of the contour are now imagined to recede to infinity, as shown in Figs. 3.9 and 3.10. One of the contours, e.g., about the branch points $u = \phi \pm i\alpha$, is selected and denoted by C. Thus, the function U_1 is given by

$$U_1 = \left(\frac{1}{4\pi}\right) \int_C \frac{g\, du}{1 - \exp i(\phi' - u)/2} \cdot \frac{(\exp ikD_1)}{D_1}. \qquad (3.14)$$

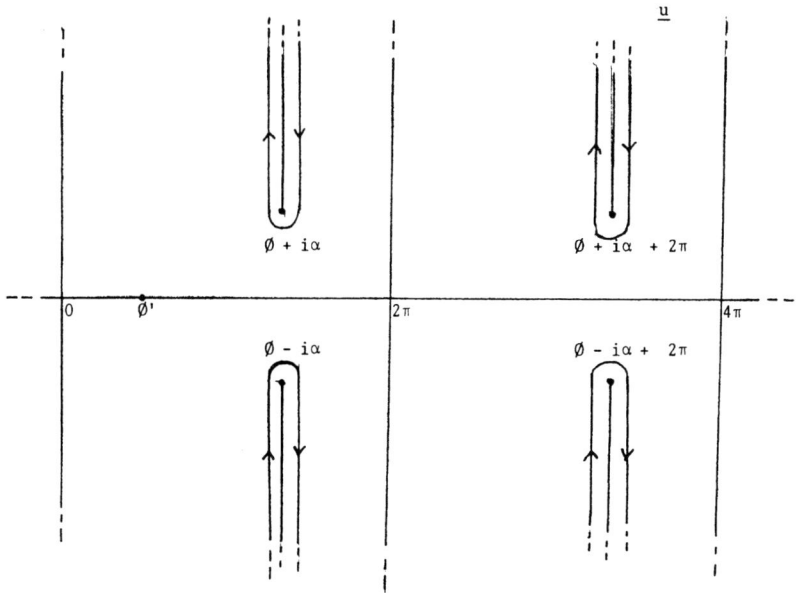

Figure 3.9 Contour equivalent to contour C_1 with horizontal sections sent to infinity $(\alpha \neq 0)$.

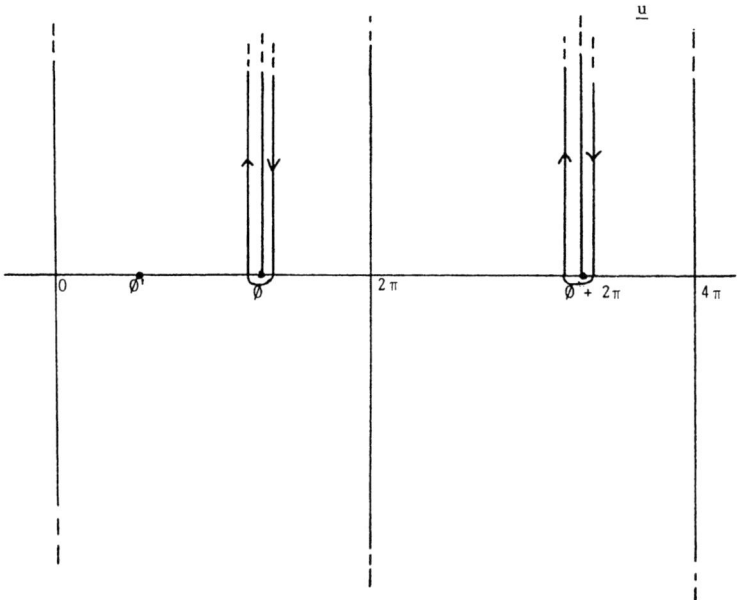

Figure 3.10 Contour equivalent to a small circle contour with horizontal sections sent to infinity $(\alpha = 0)$.

The contour C is depicted in Figs. 3.11 and 3.12 for the case $\alpha \neq 0$; a similar discussion for the case $\alpha = 0$ results in the contour shown in Fig. 3.13.

The function U_1 has the following properties:

(1) U_1 is a solution of the wave propagation equation, Eq. (3.1), because the integrand in Eq. (3.14) is also a solution of Eq. (3.1) and differentiation "under the integral sign" is permitted since the integral converges uniformly (see Appendix D). The apparent dependence of the contour on ϕ for small alterations of ϕ can be removed by the deformation of contour C.

(2) U_1 is periodic as a function of ϕ with period 4π; therefore, it is single-valued in the two-leaved space bounded by the z-axis and the fixed point P'.

(3) As $D(P, P')$ increases without bound, U_1 approaches zero as a limit in both leaves of the space. This follows upon noting that when P approaches infinity, it is equivalent to requiring that $\cosh \alpha$ approach infinity. Referring to the integral in Eq. (3.14) and passing to the limit under the integral sign, it is shown that for given $|u|$, $\cosh \alpha$ will eventually exceed the given value of $|u|$ and the value of $|D_1|$ will approach infinity. Since $\exp ikD_1$ remains bounded and the denominator increases without bound, the conclusion follows. Derivation of the real integral for U_1 in the next section verifies this conclusion.

(4) Since $U_1\left(\eta, \phi, z, P'\right) + U_1\left(\eta, \phi, +2\pi, z, P'\right)$

$$= \left(\frac{1}{4\pi}\right) \oint du \left[\frac{1}{1 - \exp i\left(\phi' - u\right)/2} + \frac{1}{1 + \exp i\left(\phi' - u\right)/2} \right] \frac{\left(\exp ikD_1\right)}{D_1},$$

$$(3.15)$$

$$= \frac{1}{4\pi} \oint du \left[\frac{1}{1 - \exp i\left(\phi' - u\right)/2} \right] \frac{\left(\exp ikD_1\right)}{D_1} = \frac{\left(\exp ikD\right)}{D}, \qquad (3.16)$$

it follows that $U_1 - (\exp ikD)/D$ is finite at the point $P = P'$. Then U_1 behaves like $(\exp ikD)/D$ in the neighborhood of P' in the first leaf (physical space) of the two-leaved space and is finite in the second leaf.

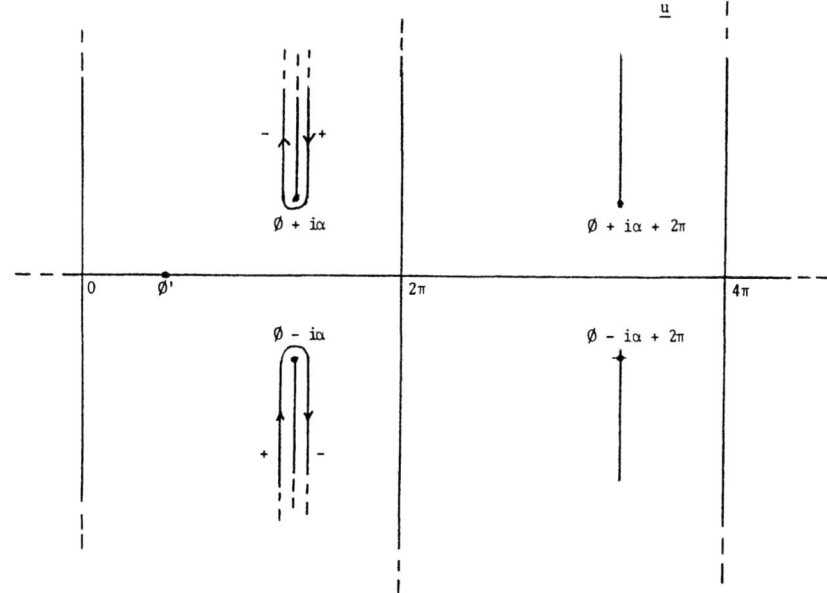

Figure 3.11 Contour C defining U_1 $(\alpha \neq 0)$. The algebraic sign next to each part of the contour indicates the sign before the radical in D_1.

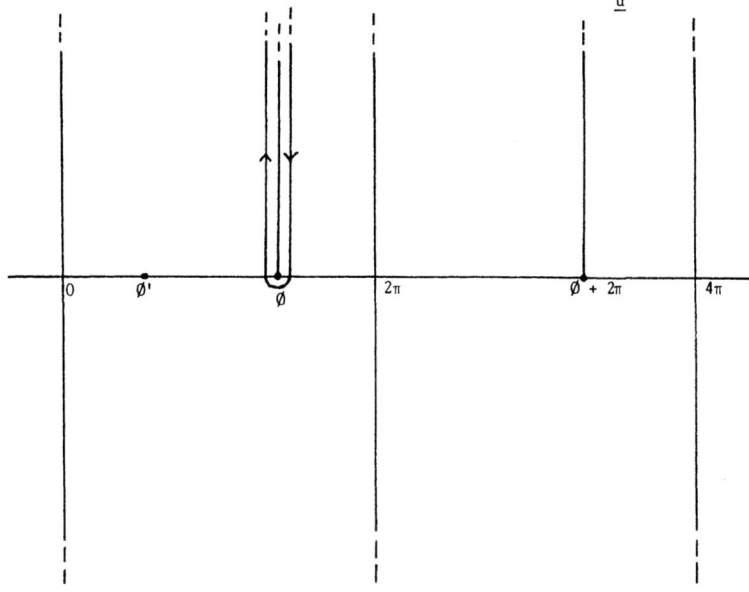

Figure 3.12 Contour equivalent to C $(\alpha = 0)$.

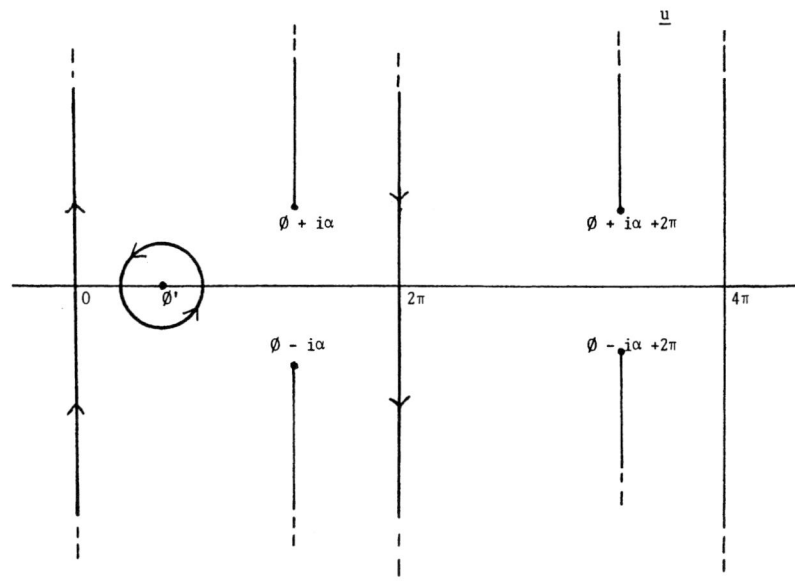

Figure 3.13 Contour C defining U_1 $(\alpha \neq 0)$.

In view of the properties (1) through (4), U_1 is termed a two-leaved generalization of the point-radiation source in physical space. Moreover, it is noted that U_1 approaches zero as $1/D$ for large values of D. This is especially relevant to the case of a branch curve that extends to infinity since it has been established that a solution with a higher degree of singularity at the branch curve cannot represent reality.[5]

3.6 Uniqueness of the Solution

Having derived a solution of the wave propagation equation defined on a two-leaved space, there is naturally a question as to how unique that solution might be. For the case of $k = 0$, there is a uniqueness theorem for solutions of Laplace's equation defined on a multileaved space from which a fixed point has been removed (i.e., a potential solution).[6]

Although a uniqueness theorem for solutions of the wave propagation equation defined on a multileaved space analogous to the one in Ref. [6] does not exist, there is a theorem for uniqueness in physical space that requires a solution V not only to approach zero like $1/D(P,O)$ as P approaches infinity (O denotes the origin of the coordinate system), but also that $D(P,O)V$ must remain finite. In addition, the radiation generated at the source point P' must diverge from the source point at a great distance from the source so that[7]

$$\lim_{D(P,O)\to\infty} D(P,O)\left[\left(\frac{\partial V}{\partial D}\right) - ikV\right] = 0 \,. \tag{3.17}$$

The condition that $D(P,O)$ increase without bound is the equivalent to the condition that $D(P,P')$ increase without bound. In any case, it is evident that the solution U_1 satisfies the above conditions in leaf 1, i.e., in physical space as well as in leaf 2. It is noted that the plane wave $\exp[ikD(P,O)]$ satisfies Eq. (3.17) since that expression is zero for all points P in ordinary space.

A further condition has been specified in view of the possibility that the "knife edge" at the branch curve (defined later as the edge of a diffracting obstacle) may contribute a singularity to the solution.[5] This possibility does not occur if any singularity on the branch curve is of the order of $1/D(P,O)$ or lesser order. This condition satisfies Sommerfeld's original solution to the Fraunhofer diffraction. For the present solution, if $\eta \to +\infty$ in the integrand of Eq. (3.14), it can be seen that the latter condition is satisfied. Interchange of the order of taking the limits in Eq. (3.14) is justified in Appendix D.

It follows, at least for physical space, that the solution U_1 satisfies the conditions for uniqueness.

3.7 Explicit Expressions for U_1

Although the integrals for U_1 for various contours cannot be evaluated explicitly in closed form in every case, approximate expressions with practical value can be derived similarly to the one used in Ref. [9].

An exception to the latter limitation occurs for $k = 0$ (i.e., an infinite wavelength) for which the function U_1 reduces to the multiple-valued generalization of Green's function defined on an infinite region bounded by one branch line, and is a solution of Laplace's equation.[4,6] Denoting this function by κ, it is found that[4]

$$\kappa(P,P') = \left(\frac{1}{D}\right)\left[\left(\frac{1}{2}\right) + \left(\frac{1}{\pi}\right)\arcsin\frac{\cos(\phi-\phi')/2}{\cosh(\alpha/2)}\right]. \tag{3.18}$$

If $k \neq 0$, an integral for U_1 is readily derived. Upon shrinking the contour C down to the branch cut barriers and noting that the contribution of the small loop about each branch point is zero in the limit, the integral for U_1 has the form

$$U_1 = \left(\frac{1}{\pi}\right) \Re \int_\alpha^\infty i \, d\eta'' \frac{\exp - ka\sqrt{2}\left\{\left[\exp - (\eta + \eta')/2\right]\sqrt{\cosh \eta'' - \cosh \alpha}\right\}}{a\sqrt{2}\left[\exp - (\eta + \eta')/2\right]\sqrt{\cosh \eta'' - \cosh \alpha}}$$
$$\times \frac{\exp - ka\left[\cos(i\eta'' - \phi' + \phi)\right]}{1 - \exp \, i(\phi' - i\eta'' - \phi)/2}.$$
(3.19)

The real part of Eq. (3.19) varies as $(\cos kD)/D$ near P'.

The first exponential factor in the numerator of the integrand varies from unity, for $\eta'' = \alpha$, to zero for $\eta'' = +\infty$; the variation is monotonic. If α is a large number, i.e., the point of observation is either close to or far from the branch line, there cannot be much variation in the latter factor. There is corresponding justification for replacing the factor with its average value. Similar remarks apply to the second exponential factor in the numerator. Having done this, the remaining integral is equal to Eq. (3.18), which provides, together with the averaged factors, an approximation for U_1.

3.8 Multiple-Valued Generalization of a Plane Wave

A fundamental concept in the theory of diffraction and scattering is the initial condition at infinity specified by an incident plane wave, i.e., Fraunhofer diffraction. Indeed, this was the subject of Sommerfeld's original paper.[10] Although the plane wave $\exp i(\mathbf{k} \cdot \mathbf{R} - \omega t)$, with $R = D(P,O)$ and $\omega = ck$, is a solution of the wave propagation equation, it is exceptional in that it does not vanish as $1/R$. However, it does satisfy the uniqueness condition discussed in Sec. 3.6.

On the other hand, in the analyses of physical optics, it is sometimes convenient to imagine that a plane wave is a limiting case for a spherical wave, observed very far from its source: $[D(P,O)/D(P',O)] \ll 1$ (see Fig. 3.14); equivalently, $(R/R' \ll 1)$. The latter physical model can be incorporated into a limiting process applied to a spherical wave representation in physical space (i.e., a single-leaved space),[11] and will be similarly employed in this section for a two-leaved space.

Thus, the spherical wave solution in physical space, $(\exp ikD)/D$, can be approximated for the case $(R/R') \ll 1 (R = +\sqrt{r^2 + z^2}, \ R' = +\sqrt{r'^2 + z'^2})$ by

$$\left.\begin{array}{c}\left[\dfrac{(\exp ikD)}{D}\right] \approx \left[\dfrac{(\exp ikR')}{R'}\right]\exp i\mathbf{k}\cdot\mathbf{R} \\[2ex] R'\left[\dfrac{(\exp ikD)}{D}\right] \approx \exp i\mathbf{k}\cdot(\mathbf{R}-\mathbf{R}')\end{array}\right\}, \qquad (3.20)$$

where $\mathbf{k}=k\hat{n}$ (Fig. 3.14). Therefore, it is expected that

$$\lim_{R'\to\infty}\left[R'\exp\left(-ikR'\right)\right]\left[\frac{\exp\left(ikD\right)}{D}\right]=\exp i\mathbf{k}\cdot\mathbf{R} \qquad (3.21)$$

in physical space.

Correspondingly, if the multiple-valued generalization of the plane wave is denoted by V_1 and of the spherical wave $(\exp ikD)/D$ by U_1, it is expected that (note that $D\approx R'$ for large R')

$$V_1 = \lim_{R'\to\infty}\left[R'\exp\left(-ikR'\right)\right]U_1\left(\eta,\phi,z,P'\right), \qquad (3.22)$$

$$= \exp i\mathbf{k}\cdot\mathbf{R} - \lim_{R'\to\infty}\left[R'\exp\left(-ikR'\right)\right]U_1\left(\eta,\phi+2\pi,z,P'\right), \qquad (3.23)$$

provided the limits exist. In any case, the approximate forms in Eq. (3.20) can be employed.

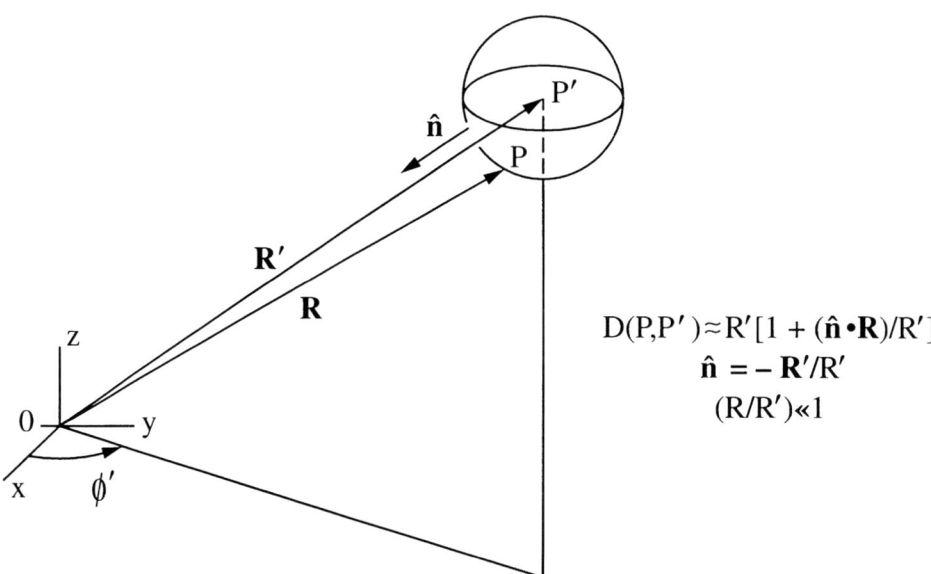

$$D(P,P')\approx R'[1 + (\hat{\mathbf{n}}\bullet\mathbf{R})/R']$$
$$\hat{\mathbf{n}} = -\mathbf{R}'/R'$$
$$(R/R')\ll 1$$

Figure 3.14 Spherical wave incident on semi-infinite plane $\left(\phi = 0\right)$. (Not to scale.)

It also follows that if $R/R' \ll 1$,

$$R'U_1 \approx \exp i\mathbf{k} \cdot (\mathbf{R} - \mathbf{R}') - R'U_1(\eta, \phi + 2\pi, z, P'). \tag{3.24}$$

Thus, for the function U_1 of this chapter, assuming that $\phi - \pi/2$ and noting that for large R', $\cosh \alpha \approx \cosh \eta'$, where $\eta' < 0$,

$$R'U_1 \approx \cos \mathbf{k} \cdot (\mathbf{R} - \mathbf{R}')$$

$$-\left(\frac{1}{\pi}\right) \Re \int_{-\eta'}^{\infty} d\eta'' \frac{\exp\left[\left(-ka\sqrt{2}\right)(\exp-\eta'/2)\right]\sqrt{\cosh \eta'' - \cosh \eta'}}{a\sqrt{2}(\exp-\eta'/2)\sqrt{\cosh \eta'' - \cosh \eta'}} \tag{3.25}$$

$$\times \frac{\exp-ka\left[\sin\left(i\eta'' + \phi\right)\right]}{1 + \frac{1-i}{\sqrt{2}}\exp\left(\frac{\eta'' - i\phi}{2}\right)}.$$

It is of interest to ask if there is a limiting process corresponding to that above for the case $k = 0$, since such functions have been well investigated. According to Ref. [6], for branch curves that do not extend to infinity (e.g., the z-axis of the present two-leaved space *does* extend to infinity), the function defined by

$$\lim_{R' \to \infty} R'\kappa(P, P') \tag{3.26}$$

is called the "harmonic measure" of the two-leaved space. [11,12] It is a solution of Laplace's equation in the two-leaved space and approaches unity as R increases without bound in one leaf of the space, and approaches zero as R increases without bound for P in the other leaf of the space. The latter theorem does not apply to the space of the present chapter; approaching infinity in any leaf would imply approaching the branch curve bounding the space: a continuum of singular points. Indeed, it is noted that as R increases without bound for P in either leaf, κ approaches $1/2D$ and the expression in Eq. (3.26) approaches a constant. However, if the branch curve is a circle, Eq. (3.26) approaches a function with the required properties of the harmonic measure (see Chapter 5).

References

1. F. Jenkins and H. White, *Fundamentals of Optics*, *Third Edition*, p. 353, McGraw-Hill, New York (1957).

2. A. Sommerfeld, "Über verzweigte Potentiale im Raum," *London Math. Soc. Proc.*, **28**, pp. 395–429 (1897).

3. S. Neustadter, "Multiple-valued harmonic functions with circle as branch curve," *Univ. Calif. Publ. in Math.*, **1**, pp. 396–432 (1951).

4. F. Alzofon, *Multiple-Valued Functions in Three-Dimensional Space and Sommerfeld's Method*, Lockheed Electronics Co., Houston, TX (1970).

5. M. Born and E. Wolf, *Principles of Optics, Sixth Edition*, pp. 591–592, Pergamon Press, New York (1980).

6. G. Evans, "Lectures on multiple-valued harmonic functions in space," *Univ. Calif. Publ. in Math.*, **1**, pp. 396–432 (1951).

7. A. Sommerfeld, *Partial Differential Equations in Physics*, Academic Press, New York (1949).

8. J. Stratton, *Electromagetic Theory*, McGraw-Hill, New York (1941).

9. A. Sommerfeld, "Mathematische Theorie der Diffraction," *Math. Ann.*, **47**, pp. 317–374 (1896).

10. J. Jackson, *Classical Electrodynamics, Second Edition*, John Wiley & Sons, New York (1975).

11. R. Nevanlinna, *Eindeutige Analytische Funktionen*, Verlag von Julius Springer, Berlin (1936).

12. N. Wiener, "Certain notions in potential theory," *J. Math. Phys.*, **3**, pp. 24–51 (1924).

Chapter 4
Fresnel Diffraction by a
Semi-Infinite Plane

In this chapter, the function U_1 derived in Chapter 3 is applied to the diffraction of a monochromatic spherical wave by a semi-infinite plane; the analysis is first applied to a scalar wave, showing that the use of the multiple-valued function U_1 as part of a Huygens' construction permits a simple (though approximate) way of accounting for the diffraction induced by a boundary condition on a semi-infinite plane. The scalar U_1 is then applied to Fresnel diffraction by a semi-infinite, perfectly conducting plane in a manner similar to Sommerfeld's analysis of Fraunhofer diffraction, and also valid for all wavelengths of the incident radiation.[1] The chapter concludes with an alternate method of analysis dependent on a suitable choice of a generator vector, valid for all wavelengths of the incident radiation.

4.1 Scalar Theory

Although a complete solution of a boundary value problem for the electromagnetic field must provide electric and magnetic vectors, calculation of these can be made to depend on the determination of a scalar field. Further justification for the study of scalar field representations of the electromagnetic field lies in the early years of the wave theory of optics when the light field was represented by a scalar function, a solution of Eq. (3.1). Additionally, in a large number of optical analysis problems, "an approximate description in terms of a single complex scalar wave function is adequate."[2] Thus, if W represents the scalar amplitude of the radiation field, $|W|^2$ can sometimes be assumed to be approximately proportional to the energy density of the field. Then, $c|W|^2$ is proportional to the energy current density, where c represents the speed of light.

Moreover, with the aid of the Huygens-Fresnel principle,[2] one can obtain an approximate description of how a given scalar wave is diffracted by obstacles and apertures. Thus, according to Huygens, "each element of a wavefront may be regarded as the center of a secondary disturbance that gives rise to spherical wavelets...the position of the wavefront at any later time is the envelope of all such wavelets."[2] Although approximate, considerable insight into light propagation and diffraction can be gained by means of Huygens' construction[3,4] to which Fresnel added the postulate that secondary wavelets interfere with one another.[2] The Kirchhoff integral, a later refinement, becomes useful only if certain assumptions are made about the value of the scalar wavefront on the diffracting surface or on the aperture over which the integral is evaluated.

Even the Kirchhoff integral can be simplified if the Green's function for the surface of integration is known, but this has been possible only in the simplest of cases, for example, a plane.[2] The Sommerfeld method, however, considerably extends the number of surfaces for which such a Green's function can be constructed.

Similar to the construction of Green's function for a region bounded by a plane for Laplace's equation $k = 0$, as discussed in Chapter 2, it is shown that Green's function for a semi-infinite plane can be constructed from the two-leaved generalization of the point radiation source U_1. That is, the Green's function for the semi-infinite plane is a function that vanishes on the semi-infinite plane, is a solution to the equation of wave propagation, and behaves essentially as (cos $kD)/D$ near the source point. An example of Huygens' construction leading to a Green's function is shown in the next section in connection with the analysis of the spherical wave's reflection by a perfectly reflecting semi-infinite plane.

One can imagine the use of the generalization U_1 to construct a generalized plane wave as an envelope of such functions with source points distributed over a plane. Such a construction would have the limitations of Huygens' method.

4.1.1 Reflection of a spherical wave by a perfectly reflecting semi-infinite plane: scalar theory

This section presents an analysis of the reflection of a spherical wave by a Huygens' construction using the multiple-valued function U_1. This approach allows a simple way of evaluating the diffraction of the spherical wave by the semi-infinite plane.

Suppose that a multiple-valued spherical wave $U_1(\eta,\phi,z,\eta',\phi,z')$ is incident on a perfectly reflecting semi-infinite plane specified by $\phi = 0$ in physical space $(0 \le \phi < 2\pi)$. The source point P' also lies in physical space. According to Huygens' construction, an incident spherical wave generates another spherical wave at the surface of the reflector equivalent to a virtual source at the mirror image of the original source. If this notion is applied to a two-leaved space, it would imply imagining another source at the image source point $P'(\eta',4\pi-\phi',z')$. The total field (taking into account a phase shift of π radians according to Stokes[4]) is then

$$W = U_1(\eta,\phi,z,\eta',\phi',z') - U_1(\eta,\phi,z,\eta',4\pi-\phi',z'). \qquad (4.1)$$

For $\phi = 0$, and owing to the appearance of ϕ' in Eq. (3.19) in the term $\cos(\phi-\phi')/2$, it is seen that $W = 0$ as desired.

Alternatively, recalling that the purpose of the reflection is to obtain a value of zero for the field on the semi-infinite plane, an equivalent substitution (insofar as the boundary condition is concerned) of $\phi \to 4\pi - \phi$ can be made to obtain

$$W = U_1(\eta,\phi,z,\eta',\phi',z') - U_1(\eta,4\pi-\phi,z,\eta',\phi',z'), \qquad (4.2)$$

and again it is clear that if $\phi = 0$, then $W = 0$. The latter is the reflection in ϕ used by Sommerfeld,[1] in contrast to the reflection of a source in the Kelvin image method.

Both of the above equations describe the same field in physical space, although one is generated by one source point and the other by two sources. Equation (4.1) is consistent with Huygens' model of reflection and Eq. (4.2) is not. However, Eq. (4.2) is a difference of Green's functions for the two-leaved space.

Owing to Eq. (3.16), one can set

$$W = \left[\frac{(\cos kD)}{D}\right] - U_1(\eta,\phi+2\pi,z,P') - U_1(\eta,-\phi,z,P'), \qquad (4.3)$$

explicitly exhibiting an incident spherical wave and its modification by the requirement of a boundary condition on the semi-infinite plane, i.e., diffraction. The density of the resultant field is proportional to W^2, and the incident energy density is proportional to $[(\cos kD)/D]^2$. The ratio of resultant intensity I and incident intensity I_o is equal to $W^2/[(\cos kD/D)]^2$. As can be seen from Eq.

(4.1), the resulting ratio is quite complex, even for this relatively simple configuration and boundary condition.

4.1.2 Diffraction of a spherical wave by a nonperfectly reflecting semi-infinite plane

Sommerfeld's original analysis[1] applied to a perfectly conducting surface and the above discussion dealt with a perfect reflector. In both cases, the analysis depended on a difference of multiple-valued functions reducing to zero on the diffracting surface. It can be shown, however, that the same kind of boundary condition can be applied to a nonperfect reflector, for example.

To this end, suppose the second leaf of a two-leaved space filled with a nonabsorbing dielectric. Associated with its boundary at $\phi = 2\pi$ is a reflectivity denoted by r and a transmissivity t such that $r + t = 1$.

A radiation point source at $P'(\eta', \phi, z')$, with $y' < 0$ in the first leaf gives rise to a spherical wave $(\cos kD)/D$ incident on the semi-infinite plane $\phi = 2\pi$. No new source is introduced by this device.

In addition to the reflected wave, a refracted wave must exist. According to Edser[3] the refracted wave originates from the virtual point source $P_t'(\eta_t', \phi_t', z_t')$, where $y_t' = my', y' = a\exp(-\eta')\sin\phi', y_t', = a\exp(-\eta_t')\sin\phi_t'$, and m denotes the index of refraction of the dielectric. It is readily shown that $\tan\phi_t' = m\tan\phi'$. In addition, it is remarked that the refracted wave, in accord with Huygens' construction, is not spherical, but is assumed to be for simplicity.[3]

Since the difference of the phases of the incident (first leaf) and refracted (second leaf) waves must be zero on the boundary $\phi = 2\pi$ $(2\pi \le \phi < 4\pi)$, it is written for the resultant transmitted wave motion that

$$t\left[U_1\left(\eta, \phi, z, \eta_t', \phi_t', z_t'\right) - U_1\left(\eta, 4\pi - \phi, z, \eta', \phi', z'\right) \right.$$
$$\left. - U_1\left(\eta, 4\pi - \phi, z, \eta_t', \phi_t', z_t'\right) + U_1\left(\eta, \phi, z, \eta', \phi', z'\right) \right]. \tag{4.4}$$

The expressions above for incident, reflected, and refracted wave motion refer to functions defined on infinite two-leaved spaces based on a Huygens' construction similar to that used for a single-leaved space. Like the expressions derived by the Kelvin image method discussed in Chapter 2, the ones derived in this section are physically relevant for only part of the region upon which they are derived, i.e., in the above examples, for one leaf.

4.2 The Electromagnetic Field Equations

A rigorous treatment of diffraction requires that radiation be treated as an electromagnetic field. This field is described by a pair of vectors, where $\mathbf{E}$ is the electric field vector measured in electrostatic units and $\mathbf{H}$ is the magnetic field vector measured in electromagnetic units, i.e., the Gaussian system of units. These vectors are related by the Maxwell equations[2]

$$\nabla \times \mathbf{H} = \left(\frac{1}{c}\right) \partial \mathbf{E}/\partial t, \tag{4.5}$$

$$\nabla \times \mathbf{E} = -\left(\frac{1}{c}\right) \partial \mathbf{H}/\partial t. \tag{4.6}$$

If there are no electric charges or currents in the region, it can be shown that both of these vectors satisfy the wave equation

$$\left(\Delta - \frac{1}{c^2} \cdot \frac{\partial^2}{\partial t^2}\right) \mathbf{V} = 0, \tag{4.7}$$

where the vector $\mathbf{V}$ may denote any one of the vectors in Eq. (4.5), for example.

The energy density w of the field may be taken to be[2]

$$w = \frac{\left(E^2 + H^2\right)}{8\pi} \left(\text{ergs cm}^{-3}\right), \tag{4.8}$$

and the energy-current density $\mathbf{S}$ can be considered equal to

$$\mathbf{S} = \left(\frac{c}{4\pi}\right)(\mathbf{E} \times \mathbf{H}) \left(\text{ergs sec}^{-1} \text{ cm}^{-2}\right). \tag{4.9}$$

For the present purpose, only monochromatic radiation will be discussed. Consequently, in the equations of motion of the electromagnetic field, the field vectors are replaced by their Fourier transforms $\mathbf{E}_\omega$ and $\mathbf{H}_\omega$; these may be complex, i.e., with real and imaginary parts. In addition, since measurements of the radiation field are carried out in terms of averages over many cycles of the oscillating field, the observed quantities w and $\mathbf{S}$ must be replaced by averages over the time, denoted by the symbols $\langle$ and $\rangle$. Thus,[2]

$$\langle w \rangle = \left(\frac{1}{16\pi}\right) \left\langle \left(\mathbf{E}_\omega \cdot \mathbf{E}_\omega^*\right) + \left(\mathbf{H}_\omega \cdot \mathbf{H}_\omega^*\right) \right\rangle \tag{4.10}$$

and

$$\left\langle \mathbf{S} \right\rangle = \left(\frac{c}{16\pi} \right) \left\langle \left[\left(\mathbf{E}_\omega \times \mathbf{H}_\omega^{\;*} \right) + \left(\mathbf{E}_\omega^{\;*} \times \mathbf{H}_\omega \right) \right] \right\rangle , \qquad (4.11)$$

where, for example $\left(\mathbf{E} = 0 \text{ if } t < 0 \right)$,

$$\mathbf{E} = \frac{1}{\sqrt{2\pi}} \int_{-\infty}^{\infty} d\omega \mathbf{E}_\omega \exp\left(-i\omega t \right),$$

$$\mathbf{E}_\omega = \frac{1}{\sqrt{2\pi}} \int_{0}^{+\infty} dt' \mathbf{E}(t') \exp\left(-i\omega t' \right). \qquad (4.12)$$

For brevity, the ω subscripts are deleted.

4.3 Boundary Conditions

For a perfect conductor, the boundary conditions may be taken to be[2]

$$E_t = 0 \qquad (4.13)$$

and

$$H_n = 0 , \qquad (4.14)$$

where E_t denotes the component of the electric vector tangential to the surface of the conductor, and H_n denotes the component of the magnetic vector normal to the conductor. These components are not necessarily Fourier transforms, but may be chosen to be. Alternatively, for the semi-infinite plane of this chapter,

$$E_x = E_z = H_y = 0 . \qquad (4.15)$$

4.4 Poincaré/Sommerfeld Solution

In this section, the boundary problem for a spherical wave incident on a perfectly conducting semi-infinite plane is solved. First, the representation of the electromagnetic field is simplified according to a suggestion credited to H. Poincaré.[1,7] Then the multiple-valued solution U_1 is used in a manner similar to the discussion in Ref. [1] to provide a solution of the boundary value problem.

Poincaré's resolution of dividing the electromagnetic field into two component fields has been applied to boundary value problems that are essentially two-dimensional (see examples in Refs. [1] and [2]). This requires a plane wave representation independent of the z-coordinate in Ref. [1]. The spherical wave generalization U_1, however, is not only dependent on η and ϕ, but also on z, as defined in a three-dimensional space. To avoid the latter difficulty, imagine all the field quantities to be Fourier analyzed in the z-coordinate. The following analysis is then concerned with Fourier transforms in both the time and the z-coordinate: indices to indicate the transformation are deleted, as well as the factors $\exp i\omega t$ and $\exp ihz$. Where a derivative with respect to z has occurred in the original equations, it is replaced by the factor ih.

Thus, Eqs. (4.5) and (4.6) can be written as $\left(k=\omega/c\right)$

$$\left.\begin{array}{c} \left(\dfrac{\partial E_z}{\partial y}\right) - ih\,E_y = ik\,H_x \\[2mm] ih\,E_x - \left(\dfrac{\partial E_z}{\partial x}\right) = ik\,H_y \\[2mm] \left(\dfrac{\partial H_y}{\partial x}\right) - \left(\dfrac{\partial H_x}{\partial y}\right) = -ik\,E_z \end{array}\right\} \qquad (4.16)$$

and

$$\left.\begin{array}{c} \left(\dfrac{\partial H_z}{\partial y}\right) - ih\,H_y = -ik\,E_x \\[2mm] ih\,H_x - \left(\dfrac{\partial H_z}{\partial x}\right) = -ik\,E_y \\[2mm] \left(\dfrac{\partial E_y}{\partial x}\right) - \left(\dfrac{\partial E_x}{\partial y}\right) = ik\,H_z \end{array}\right\} . \qquad (4.17)$$

In accordance with Ref. [6] for "E-polarization,"

$$E_x = E_y = H_z = 0, \qquad (4.18)$$

and Eqs. (4.16) become

$$\left.\begin{array}{c} \left(\dfrac{\partial E_z}{\partial y}\right) = ik\,H_x \\[2mm] -\left(\dfrac{\partial E_z}{\partial x}\right) = ik\,H_y \\[2mm] \left(\dfrac{\partial H_y}{\partial x}\right) - \left(\dfrac{\partial H_x}{\partial y}\right) = -ik\,E_z \end{array}\right\} \qquad (4.19)$$

so that

$$\left[\left(\frac{\partial^2}{\partial x^2}\right)+\left(\frac{\partial^2}{\partial y^2}\right)+k^2\right]E_z=0 . \qquad (4.20)$$

It follows that all the nonzero field components can be determined from E_z.

Similarly, for "H-polarization,"

$$H_x=H_y=E_z=0 ; \qquad (4.21)$$

and Eqs. (4.17) become

$$\left.\begin{array}{r}\left(\dfrac{\partial H_z}{\partial y}\right)=-ik\ E_x \\[2mm] \left(\dfrac{\partial H_z}{\partial x}\right)=ik\ E_y \\[2mm] \left(\dfrac{\partial E_y}{\partial x}\right)-\left(\dfrac{\partial E_x}{\partial y}\right)=ik\ H_z\end{array}\right\}, \qquad (4.22)$$

from which it follows that

$$\left[\frac{\partial^2}{\partial x^2}+\frac{\partial^2}{\partial y^2}+k^2\right]H_z=0 \qquad (4.23)$$

so all nonzero field components can be determined from H_z.

Now, for E-polarization,

$$E_z=U_{1h}\left(\eta,\phi,z,P'\right)-U_{1h}\left(\eta,4\pi-\phi,z,P'\right), \qquad (4.24)$$

where the notation U_{1h} denotes the Fourier transform of U_1 in z. It follows that E_z = 0 if $\phi=0$, and since E_x = 0 on the half-plane, the boundary condition E_t = 0 is satisfied. Because E_z is zero everywhere on the semi-infinite plane and therefore cannot vary with x, H_y = H_n = 0 on this surface—a conclusion that follows from the second part of Eqs. (4.17). Thus, all of the boundary conditions are satisfied.

The reasoning for H-polarization is very similar, except that now H_z is the sum of the two terms in Eq. (4.24) and the derivative, with respect to y on the semi-infinite plane, is equal to the derivative with respect to ϕ for ϕ = 0. It is then readily seen that the boundary conditions are all satisfied.

4.5 Solution Using Two Independent Scalar Solutions

The following method of solution was motivated by analyses in Refs. [5] and [8] and presents an alternative method of solution to the boundary value problem discussed in Sec. 4.4.

In essence, the method consists of choosing two independent solutions v and w of the scalar equation of wave propagation Eq. (3.1) and using them to construct two independent vectors that are combined to represent the electric and magnetic field vectors. It is then shown that these field vectors satisfy the boundary conditions. The geometric configuration on which the boundary conditions are imposed dictates the manner of defining the two independent solutions of Eq. (3.1) and of constructing the two independent vectors.

In the present case, the two independent solutions to the equation of wave propagation are the difference and the sum of values of the function U_1 previously introduced. Thus, for convenience, when all the variables are deleted except ϕ in the function notation for U_1,

$$v = U_1(\phi) - U_1(-\phi) \qquad (4.25)$$

and

$$w = U_1(\phi) + U_1(-\phi). \qquad (4.26)$$

Then it is defined that

$$\mathbf{M}_\psi = \nabla \times \left(\hat{\mathbf{a}}_y \psi \right), \qquad (4.27)$$

where $\hat{\mathbf{a}}_y$ denotes a unit vector along the y-axis and ψ denotes a solution of the wave propagation equation; also defined is

$$k\mathbf{N}_\psi = \nabla \times \mathbf{M}_\psi. \qquad (4.28)$$

Then, defining

$$\mathbf{E} = \mathbf{M}_v + i\mathbf{N}_w \qquad (4.29)$$

and

$$\mathbf{H} = -\mathbf{M}_w + i\mathbf{N}_v, \qquad (4.30)$$

it is found that these vectors are the solutions of the boundary value problems.

The latter conclusion follows upon evaluating the vectors in terms of the coordinates. Thus,

$$E_x = -\left(\frac{\partial v}{\partial z}\right) + \left(\frac{i}{k}\right)\left(\frac{\partial^2 w}{\partial x \partial y}\right)$$

and

$$E_z = \frac{i}{k}\left[\left(\frac{\partial^2}{\partial z \partial y}\right)w + \frac{\partial v}{\partial x}\right],$$

where, by repetition of the reasoning in Sec. 4.4, it is shown that $E_t = 0$. Further, upon calculating H_y, it is found to be proportional to $[(\partial^2/\partial z^2) + (\partial^2/\partial x^2)]v$, which vanishes on the semi-infinite plane. The desired conclusion then follows.

If U_1 is replaced by an approximation to a plane wave (i.e., $R'U_1$ for $R/R' \ll 1$), defined on a two-leaved space as in Chapter 3, the boundary value problem of Chapter 3 corresponds approximately to Fraunhofer diffraction.

References

1. A. Sommerfeld, "Mathematische Theorie der Diffraction," *Math. Ann.*, **47**, pp. 317–374 (1896).

2. M. Born and E. Wolf, *Principles of Optics*, *Sixth Edition*, Pergamon Press, New York (1980).

3. E. Edser, *Light for Students*, MacMillan and Co., Ltd., London (1931).

4. F. Jenkins and H. White, *Fundamentals of Optics*, *Third Edition*, McGraw-Hill, New York (1957).

5. J. Stratton, *Electromagnetic Theory*, McGraw-Hill, New York (1941).

6. H. van de Hulst, *Light Scattering by Small Particles*, Dover Publications, New York (1981).

7. H. Poincaré, "Sur la polarisation par diffraction," *Acta Math.*, **16**, pp. 297–339 (1892).

8. J. Jackson, *Classical Electrodynamics*, *Second Edition*, John Wiley and Sons, New York (1975).

Chapter 5
Fresnel Diffraction by a Circular Disk

In this chapter, a representation of a spherical wave diffracted by a perfectly conducting circular disk is derived. The analysis can be adapted to a circular aperture as well as to a spherical cap.

Some of the same procedure for the solution construction already introduced for a straight-line branch curve can be carried over to this representation. The difference in the two cases can be exploited to provide a guide to generalizing the Sommerfeld method.

This discussion is typically divided into four parts: (1) the definition of a suitable coordinate system; (2) an algebraic analysis of how the branch points of the distance function explicitly depend on the point of observation, P; (3) a construction of the multiple-valued solution for a point-radiation source in the two leaved space; and (4) the application of the solution to the diffraction problem.

5.1 Coordinate-System Construction

In order to generate a coordinate system for a multileaved space bounded by a single branch circle, two line-charge sources are imagined perpendicular to the $\rho = r + iz$ -plane $(-\infty < r < +\infty)$, passing through the points $\rho = +a$ and $\rho = -a$ from which it can be written that

$$\frac{\rho - a}{\rho + a} = \left(\frac{d_1}{d_2}\right) \exp i\left(\theta_1 - \theta_2\right), \qquad (5.1)$$

where $d_1 = |\rho - a|$ and $d_2 = |\rho + a|$. The line charges are not branch curves, but a means of generating a two-dimensional coordinate system (i.e., a "toroidal" coordinate system). The symbols in Eq. (5.1) refer to the configuration illustrated in Fig. 5.1.

A point in the ρ-plane can be specified either by $\rho = a + d_1 \exp i\theta_1$ or by $\rho = -a + d_2 \exp i\theta_2$; but to describe circuits of either $+a$ or $-a$ by a single angle, the composite angle $\theta = \theta_1 - \theta_2$ is introduced.

Now, setting $(d_1 / d_2) = \exp(-\eta)$, with $-\infty < \eta < +\infty$, it can be written that

$$\frac{\rho - a}{\rho + a} = \exp(-\eta + i\theta), \qquad (5.2)$$

where $(-\pi < \theta \le +\pi)$, as illustrated in Fig. 5.2. The addition of the angle ϕ, defined as in the cylindrical coordinate system discussed in Chapter 3, with $0 \le \phi < 2\pi$, provides the coordinate system $\{(\eta, \theta, \phi)\}$. The coordinates can be expected to behave like those introduced in Chapter 3, close to either $\rho = +a$ or $\rho = -a$. Thus, if $\rho \approx a$, then $\rho - a \approx 2a \exp(-\eta + i\theta)$, which is similar to the case for a single branch line. This circumstance is a guide for further development of the Sommerfeld method.

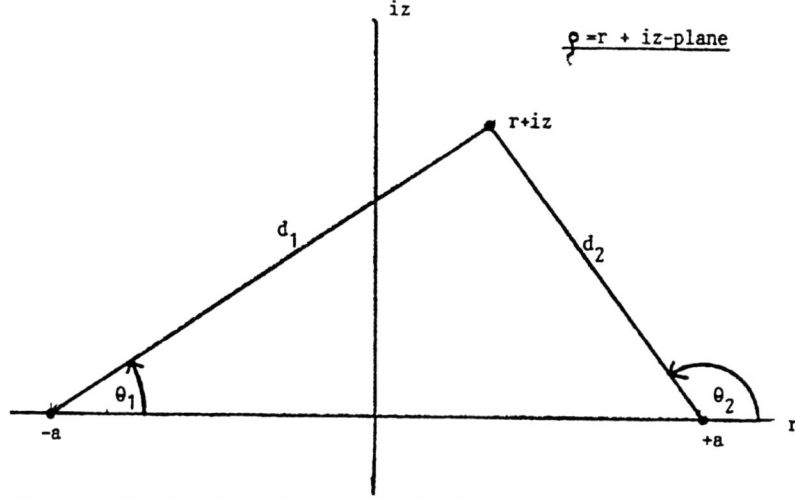

Figure 5.1 Configuration and symbols for definition of a toroidal coordinate system.

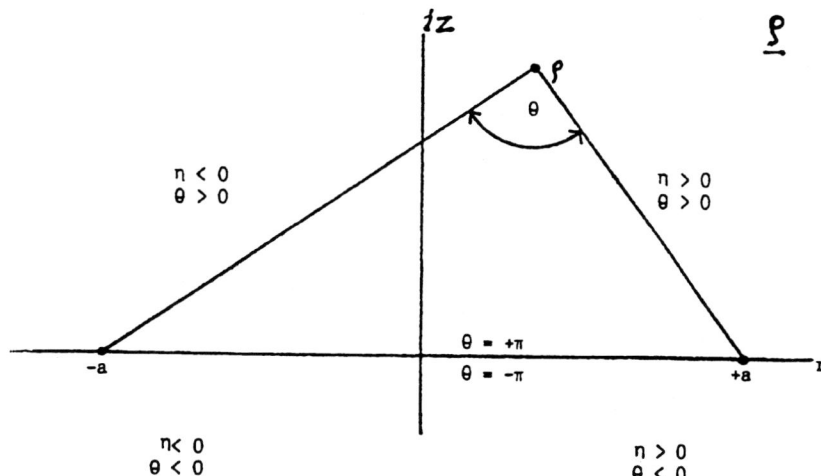

Figure 5.2 Specification of coordinate system quadrants by toroidal coordinates.

The curves $\eta = \mu = $ constant are the circles

$$\left(r - a\coth\mu\right)^2 + z^2 = a^2\mathrm{csch}^2\mu .\qquad(5.3)$$

If $\eta > 0$, the circles contain the point $\rho = +a$ in their interiors and lie to the right of the z-axis: $r > 0$. If $r < 0$, then $\eta < 0$, and the circles all contain the point $\rho = -a$ in their interiors. For the purposes of analysis, it is convenient to introduce values of r and η that are negative as well as positive, but for applications only positive or zero values of r and η will be used when coupled with the coordinate θ.

The curves $\theta = v = $ constant are the circles

$$r^2 + \left(z - a\cot v\right)^2 = a^2\csc^2 v ,\qquad(5.4)$$

all of which pass through the points $\rho = \pm a$. Arcs of these circles above the r-axis correspond to positive values of θ, and arcs of the circles below the r-axis correspond to negative values of θ. The value $\theta = +\pi$ corresponds to the upper surface of a circular disk of radius a; the value $\theta = -\pi$ corresponds to the lower surface of the disk. The interval $+\pi < \theta \leq 3\pi$ specifies leaf 2 of a two-leaved space when coupled with the other two coordinate intervals. The values $\theta = -\pi$ and $\theta = 3\pi$ are identified: they are the same surface. The circle $r = a$ is a branch curve bounding the two leaves of the space. Some of these properties are illustrated in Fig. 5.3.

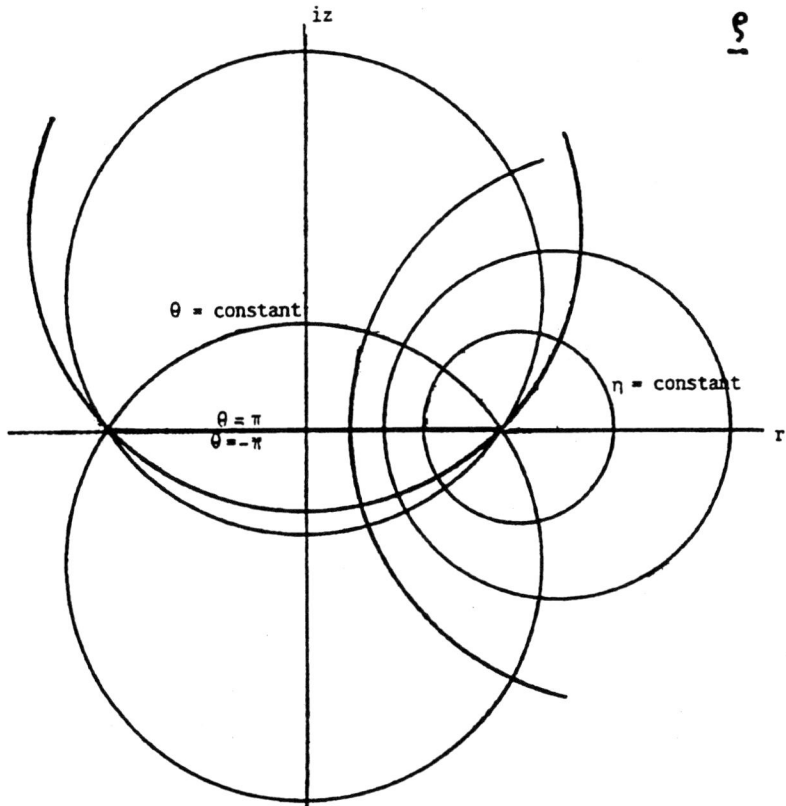

Figure 5.3 Some equipotentials and lines of force for a coordinate system generated by two line sources: the toroidal coordinate system.

It can be shown that

$$r = \frac{a \sinh \eta}{\cosh \eta - \cos \theta},$$
(5.5)

and

$$z = \frac{a \sin \theta}{\cosh \eta - \cos \theta}.$$
(5.6)

The distance function, in terms of the above coordinates, is

$$D(P,P') = \frac{a\sqrt{2}\sqrt{\cosh\alpha - \cos(\theta - \theta')}}{\sqrt{\cosh\eta - \cos\theta}\sqrt{\cosh\eta' - \cos\theta'}},$$
(5.7)

where $\cosh\alpha = \cosh(\eta - \eta') + \sinh\eta\sinh\eta'[1 - \cos(\phi - \phi')]$.

A circular hole in an infinite plane can be described mathematically by redefining the angle as shown in Fig. 5.4. The analysis to follow would not be essentially altered by this redefinition.

In addition, the angle intervals may be altered to $\theta_o - 2\pi < \theta \leq \theta_o$ for leaf 1 and $\theta_o < \theta \leq \theta_o + 2\pi$ for leaf 2, where $-\pi < \theta_0 < +\pi$. These new intervals describe a physical space bounded by a spherical cap and are suitable for analysis of the diffraction of a spherical wave by a spherical cap.

5.2 Analytic Continuation of θ'

The coordinate θ' is now continued analytically to $u = \theta'' + i\eta''$, inducing a transformation of ρ' and $\rho'*$, where

$$\rho' = a\frac{1 + \exp(-\eta' + i\theta')}{1 - \exp(-\eta' + i\theta')}, \tag{5.8}$$

so that ρ' is altered to

$$\sigma = a\frac{1 + \exp(-\eta' + iu)}{1 - \exp(-\eta' + iu)}, \tag{5.9}$$

$$= a\frac{1 + \exp(-\eta' - \eta'' + i\theta'')}{1 - \exp(-\eta' - \eta'' + i\theta'')}, \tag{5.10}$$

and $\rho'*$ becomes

$$\delta = a\frac{1 + \exp(-\eta' - iu)}{1 - \exp(-\eta' - iu)}, \tag{5.11}$$

$$= a\frac{1 + \exp(-\eta' + \eta'' - i\theta'')}{1 - \exp(-\eta' + \eta'' - i\theta'')}. \tag{5.12}$$

From Eq. (5.2) it follows that

$$\frac{\sigma - a}{\sigma + a} \cdot \frac{\delta - a}{\delta + a} = \exp(-2\eta'). \tag{5.13}$$

The analytic continuation of D results from the substitution of u for θ' in Eq. (5.7) with branch points $u_b = \theta \pm i\alpha + 2\pi p$ $(p = 0, \pm 1, \pm 2...)$ and $\pm i\eta' + 2\pi p$. The analytic continuation of D is denoted by D_2.

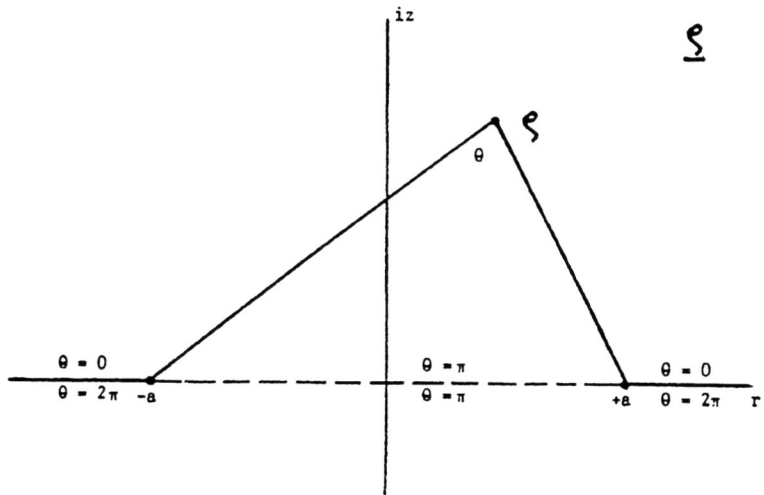

Figure 5.4 Redefinition of the angle θ to describe a screen with a circular hole.

However, for the application of the Sommerfeld method, it is undesirable to construct a Cauchy integral with branch points not dependent on θ. The latter fixed branch points can be eliminated without altering the dependence of the resulting expression on P by multiplying the analytic continuation by $\sqrt{\cosh\eta' - \cos u}\,/\sqrt{\cosh\eta' - \cos\theta'}$ to obtain

$$D_1 = \frac{a\sqrt{2}\sqrt{\cosh\alpha - \cos(u-\theta)}}{\sqrt{\cosh\eta - \cos\theta}\sqrt{\cosh\eta' - \cos\theta'}} \tag{5.14}$$

with branch points $u_b = \theta \pm i\alpha + 2\pi p\ (p = 0, \pm 1, \pm 2, ...)$, provided that the integrand of the Cauchy integral remains a solution of the wave equation.

5.3 A Multiple-Valued Green's Function with a Circle as a Branch Curve

The following construction of a Green's function for a space bounded by one branch circle essentially follows the procedure outlined in Chapter 3. However, there is a difficulty not previously encountered. Overcoming the problem will be preparation for the construction of Green's functions for more complicated cases—for example, a space bounded by two branch circles.

For the present case, the initial form of the integrand for the Cauchy integral (Sec. 3.3) has the form

$$\frac{\exp ikD_2}{D_2} = \frac{\exp\left(\dfrac{ika\sqrt{2}\sqrt{\cosh\alpha - \cos(u-\theta)}}{\sqrt{\cosh\eta - \cos\theta}\sqrt{\cosh\eta' - \cos u}}\right)}{\dfrac{a\sqrt{2}\sqrt{\cosh\alpha - \cos(u-\theta)}}{\sqrt{\cosh\eta - \cos\theta}\sqrt{\cosh\eta' - \cos u}}}. \tag{5.15}$$

The radicals containing the term $\cos(u-\theta)$ have branch points with the desired dependence on θ, i.e., $u_b^{(1)} = \theta \pm i\alpha + 2\pi p\,(p=0,\pm1,\pm2,...)$; however, the remaining branch points, $u_b^{(2)} = \pm i\eta' + 2\pi p$, are not of use in the Sommerfeld construction procedure.

If the integrand in Eq. (5.15) is multiplied by a factor independent of the coordinates of the variable point $P(\eta,\theta,\phi)$, it remains a solution of the wave equation. Thus, upon multiplication of the integrand by the factor $\sqrt{\cosh\eta' - \cos\theta'}/\sqrt{\cosh\eta' - \cos u}$, the denominator on the right side of Eq. (5.15) is transformed to D_1, defined in Eq. (5.14).

A similar multiplication cannot be applied to the exponential $\exp ikD_2$. To avoid the necessity of adding a branch cut barrier at the points $u_b^{(2)}$, one can resort to a device used by Sommerfeld in Ref. [1] [see Appendix B, Eq. (B.8)]. That is, one adds to the integrand $\exp ikD_2/D_1$ its value when either of the branch points $u_b^{(2)}$ is circuited once by the point u, changing the sign of the radical $\sqrt{\cosh\eta' - \cos u}$. Dividing the sum by 2, one obtains a new integrand: $(\cos kD_2)/D_1$. The new Cauchy integral has the value $(\cos kD)/D$, a Green's function defined in ordinary or physical space and the starting point for the Sommerfeld construction of a multiple-valued solution of the wave equation. Moreover, with the addition of branch cut barriers at the points $u_b^{(1)}$ (as in Chapter 3), the integrand of the Cauchy integral is single-valued, as required.

Indeed, the generalization to a multiple-valued solution to the wave equation is now immediate and can be written, for $D(P,P') \neq 0$:

$$U_1 = \frac{2}{\pi} \int_\alpha^\infty \frac{d\eta''}{1 - \exp\dfrac{i(\theta' - \theta - i\eta'')}{2}} \frac{\cos\dfrac{ka\sqrt{2}\sqrt{\cosh\eta'' - \cosh\alpha}}{\sqrt{\cosh\eta - \cos\theta}\sqrt{\cosh\eta' - \cos(\theta + i\eta'')}}}{\dfrac{a\sqrt{2}\sqrt{\cosh\eta'' - \cosh\alpha}}{\sqrt{\cosh\eta - \cos0}\sqrt{\cosh\eta' - \cos\theta'}}}. \tag{5.16}$$

It is noted that if P lies in the first leaf,

$$U_1(\theta) = \left[\frac{(\cos kD)}{D}\right] - U(\theta + 2\pi), \tag{5.17}$$

where $U_1(\theta+2\pi)$ has no singularities. Therefore, defining

$$U_2(\theta)=U_1(\theta)+\frac{(i\sin kD)}{D}, \tag{5.18}$$

it is found that

$$U_2(\theta)+U_2(\theta+2\pi)=\frac{(\exp ikD)}{D}, \tag{5.19}$$

so that U_2 behaves essentially as $(\exp ikD)/D$ if $D(P,P')$ is very small. Therefore, a multiple-valued generalization of the complex Green's function $(\exp ikD)/D$ is obtained, as well as one of $(\cos kD)/D$.

5.3.1 Plane Wave Approximation

For large $R'=D(P',O)$, it is noted that $R'\approx a\sqrt{2}/\sqrt{\cosh\eta'-\cos\theta'}$, and it follows that

$$R'U_1\approx\cos\left[\mathbf{k}\cdot(\mathbf{R}-\mathbf{R}')\right]-\left(\frac{2}{\pi}\right)\int_\eta^\infty\frac{d\eta''}{1+\exp(\eta''-i\theta)/2}\frac{\sqrt{\cosh\eta-\cos\theta}}{\sqrt{\cosh\eta''-\cosh\eta}}$$

$$\times\cos\left[\frac{ka\sqrt{2}\sqrt{\cosh\eta''-\cosh\eta}}{\sqrt{\cosh\eta-\cos\theta}\sqrt{1-\cosh(\theta+i\eta'')}}\right], \tag{5.20}$$

with a similar expression for U_2.

5.3.2 Static solution and the harmonic measure of the two-leaved space

If $k = 0$, the integrand of the Cauchy integral can be simplified, omitting the term $\cos kD$ and the necessity to symmetrize. The rest of the procedure is straightforward, and the resulting integral can be evaluated explicitly to have the value

$$\kappa=\frac{1}{D}\left[\frac{1}{2}+\frac{1}{\pi}\arcsin\frac{\cos(\theta-\theta')/2}{\cosh(\alpha/2)}\right], \tag{5.21}$$

a unique solution of Laplace's equation.[2] It has only one singularity at P', where it behaves essentially as $1/D(P,P')$ and approaches zero as a limit in both leaves of the two-leaved space as D increases without bound.

Analogous to the case for $k\neq0$, the limit of $D\kappa$, or $R'\kappa$, as R' increases without bound is

$$V_1 = \left(\frac{1}{2}\right) + \left(\frac{1}{\pi}\right) \arcsin \frac{\cos(\theta/2)}{\cosh(\eta/2)} . \tag{5.22}$$

The multiple-valued function V_1 is called the *harmonic measure* of the two-leaved space. It is a unique solution of Laplace's equation and tends to unity in the first leaf of the space as R increases without bound, and to zero in the second leaf of the space as R increases without bound.[2]

5.3.3 An alternative method of constructing a multiple-valued spherical wave

While the preceding method of deriving a multiple-valued spherical wave is straightforward, it is of interest to consider an alternate derivation.

The derivation presented in this section is suggested by Sommerfeld's failed construction of a Green's function solution of Laplace's equation for a space bounded by two parallel branch lines.[3] Sommerfeld introduced a bipolar coordinate system generated by two line sources (see Chapter 7), similar to the case for the toroidal coordinate system. The form of the all-important distance function D is the same as for the toroidal coordinate system, with an alteration in the definition of $\cosh \alpha$. Thus,

$$D(P,P') = \frac{a\sqrt{2}\sqrt{\cosh\alpha - \cos(\theta - \theta')}}{\sqrt{\cosh\eta - \cos\theta}\sqrt{\cosh\eta' - \cos\theta'}}, \tag{5.23}$$

where $\cosh\alpha = \cosh(\eta - \eta') + [(z - z')^2 / 2a^2](\cosh\eta - \cos\theta)(\cosh\eta - \cos\theta')$.

To obtain the desired dependence of the branch points on θ, i.e., $u_b = \theta \pm i\alpha + 2\pi p (p = 0, \pm 1, \pm 2, ...)$, Sommerfeld made an analytic continuation of θ' in the term $\cos(\theta - \theta')$ in Eq. (5.23) alone, and not in the terms $\cosh\alpha$ and $\cos\theta'$. When denoting the resulting function by D_1, it can be shown[4] that $1/D_1$ is not a solution of Laplace's equation. However, if a Cauchy integral is constructed as in Sec. 3.3 using $1/D_1$ in the integrand, the value of the integral is equal to $1/D$, which is a solution of Laplace's equation. A multiple-valued function can be constructed, nevertheless, using $1/D_1$ in the integrand and the same procedure as outlined in Sec. 3.3. The resulting function, however, is not a solution of Laplace's equation, although it does possess the other desirable features of the Green's function for the space.

The purpose of the following discussion is to show how the simplified analytic continuation employed by Sommerfeld can be used to construct a multiple-valued Green's function representing a spherical wave in physical space. The function, a solution of the wave equation, is defined on a space bounded by a branch circle. It may be that the alternate solution of this section lends itself more readily to computer calculations (a possibility that has not yet been evaluated).

Thus, returning to the function D_1 defined in Eq. (5.14) for a toroidal coordinate system, and carrying through the procedure outlined in Sec. 3.3 for the integrand of the Cauchy integral containing the expression $(\cos kD_1)/D_1$, a multiple-valued function, W_1, is obtained [expressed in Eq. (5.24) below]. Analogous to Sommerfeld's incorrect derivation, it has all of the properties desired for a Green's function (i.e., for a spherical wave in physical space), but it is not a solution of the wave equation. Also, like Sommerfeld's incorrect function, it has a simpler form than the solution derived in Sec. 5.2.

The real integral form of W_1 is given by

$$W_1 = \left(\frac{1}{4\pi}\right)\int_\alpha^\infty d\eta'' \frac{\exp\left[-ka\sqrt{2}\sqrt{\cosh\eta'' - \cosh\alpha}\big/\sqrt{\cosh\eta' - \cos\theta'}\sqrt{\cosh\eta - \cos\theta}\right]}{a\sqrt{2}\sqrt{\cosh\eta'' - \cosh\alpha}\big/\sqrt{\cosh\eta' - \cos\theta'}\sqrt{\cosh\eta - \cos\theta}}$$

$$\times \frac{\sinh(\eta''/2)}{\cosh(\eta''/2) - \cos(\theta - \theta')/2}. \tag{5.24}$$

If $k \neq 0$, then (1) W_1 is single valued in the two-leaved space bounded by a branch circle of radius a; (2) W_1 has only one singularity (in leaf 1) and $W_1 - (\cos kD)/D$ is finite everywhere in leaf 1, particularly at the point P'; (3) W_1 vanishes as $D(P, P')$ increases without bound in both leaves; and (4) the sum of the values of W_1 at underlying points is equal to $(\cos kD)/D$.

In the special case of $k = 0$, W_1 becomes the unique solution to Laplace's equation, κ, and the sum of its values at underlying points is $1/D$; in fact,

$$\kappa = \left(\frac{1}{D}\right)\left[\left(\frac{1}{2}\right) + \left(\frac{1}{\pi}\right)\arcsin\frac{\cos(\theta - \theta')/2}{\cosh(\alpha/2)}\right]. \tag{5.25}$$

Moreover, denoting the limit of $R'\kappa$, as R' increases without bound (here $R' = +\sqrt{r'^2 + z'^2} \approx a\sqrt{2}\big/\sqrt{\cosh\eta' - \cos\theta'}$, for large R', it is found that

$$V_1 = \left(\frac{1}{2}\right) + \left(\frac{1}{\pi}\right)\arcsin\frac{\cos(\theta/2)}{\cosh(\eta/2)}, \tag{5.26}$$

where V_1 is a solution of Laplace's equation. Single-valued and finite in the two-leaved space, V_1 tends to unity in leaf 1 and to 0 in leaf 2 as P tends to infinity.

The sum of the values of V_1 at underlying points is unity; V_1 is called the *harmonic measure* of the space (Sec. 5.3.2).

Returning to the discussion of $k \neq 0$, it is evident that it would not take much of an alteration to make W_1 equal to the desired kind of solution to the wave equation by the addition of a correction term. In making such a correction, it is sufficient and convenient to correct $W_1(\theta + 2\pi)$, where $-\pi < \theta \le \pi$, since $W_1(\theta) = [(\cos kD)/D] - W_1(\theta + 2\pi)$ and $W_1(\theta + 2\pi)$ has no singularity at $P = P'$. That is, W_1 need only be corrected in one leaf in order to be corrected in both leaves.

The suggested modification is given by

$$U_1(\theta + 2\pi) = W_1(\theta + 2\pi) - \left[\frac{1}{(2\pi)^{3/2}}\right] \int d\mathbf{k}'' \frac{\left[\left(\Delta + k^2\right)W_1(\theta + 2\pi)\right]_{\mathbf{k}''}}{-k''^2 + k^2} \exp(i\mathbf{k}'' \cdot \mathbf{r}),$$

(5.27)

where the Fourier transform $[(\Delta + k^2)W_1(\theta + 2\pi)]_{\mathbf{k}''}$ is given by

$$\left[\left(\Delta + k^2\right)W_1(\theta + 2\pi)\right]_{\mathbf{k}''} = \left[\frac{1}{(2\pi)^{3/2}}\right]$$
$$\times \int d\mathbf{r}'' \left(\Delta'' + k^2\right)W_1(\theta'' + 2\pi) \exp(i\mathbf{k}'' \cdot \mathbf{r}'').$$

(5.28)

The notations $d\mathbf{k}''$ and $d\mathbf{r}''$ indicate volume elements at the position vectors $\mathbf{k}''$ and $\mathbf{r}''$, respectively; the integral over the elements $d\mathbf{r}''$ is extended over leaf 2. The symbol $''$, used as a superscript for the operator Δ, indicates that the differentiation is applied to coordinates $\mathbf{r}''$, i.e., η'', θ'', z''.

It follows from Eq. (5.28) that

$$\left(\Delta + k^2\right)W_1(\theta + 2\pi) = \left[\frac{1}{(2\pi)^{3/2}}\right] \int d\mathbf{k}'' \left[\left(\Delta + k^2\right)W_1(\theta + 2\pi)\right]_{\mathbf{k}''} \exp i\mathbf{k}'' \cdot \mathbf{r},$$

(5.29)

and it is seen that applying the operator $(\Delta + k^2)$ to the right side of Eq. (5.27) leads to a value of zero; therefore, the equation of wave propagation is satisfied.

Substitution of Eq. (5.28) in Eq. (5.27) leads to the integral

$$\left[\frac{1}{(2\pi)^3}\right] \int d\mathbf{r}'' \left(\Delta'' + k^2\right)W_1(\theta + 2\pi) \int d\mathbf{k}'' \frac{\exp i\mathbf{k}'' \cdot (\mathbf{r} - \mathbf{r}'')}{k^2 - k''^2},$$

(5.30)

where the integral over k'' can be evaluated as [noting that $|\mathbf{r} - \mathbf{r}''| = D(P, P'')$]

$$\int d\mathbf{k}'' \frac{\exp i\mathbf{k}'' \cdot (\mathbf{r} - \mathbf{r}'')}{k^2 - k''^2} = 4\pi \int_0^\infty dk'' k''^2 \frac{\sin k''|\mathbf{r} - \mathbf{r}''|}{k''|\mathbf{r} - \mathbf{r}''|} \cdot \frac{1}{k^2 - k''^2}$$

$$= \frac{1}{8ki|\mathbf{r} - \mathbf{r}''|} \int_{-\infty}^\infty dk'' k'' -\left[\exp ik''|\mathbf{r} - \mathbf{r}''| - \exp-\left(ik''|\mathbf{r} - \mathbf{r}''|\right)\right] \cdot \left(\frac{1}{k'' - k} - \frac{1}{k'' + k} \right),$$

(5.31)

where the first integral is understood to be a principal value. The latter integral can be evaluated by means of the contour integrals illustrated in Fig. 5.5. It is then found for the value of the integral in Eq. (5.31) that

$$\left[\frac{i}{(4\pi)} \right] \int d\mathbf{r}'' \left(\Delta'' + k^2 \right) W_1 \left(\theta'' + 2\pi \right) \left[\frac{\sin k|\mathbf{r} - \mathbf{r}''|}{|\mathbf{r} - \mathbf{r}''|} \right].$$

(5.32)

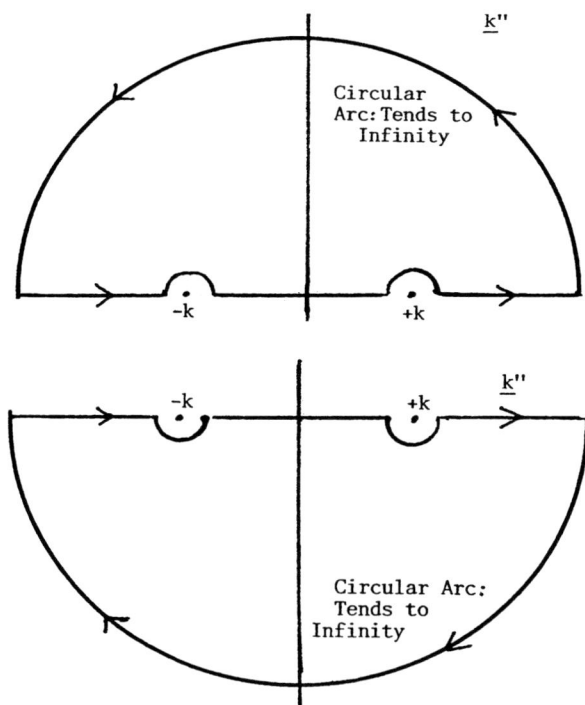

Figure 5.5 Contour in the k''-plane for evaluation of terms multiplied by $\exp(ik''|\mathbf{r} - \mathbf{r}''|)$ in the integral of Eq. (5.31).

Upon defining

$$U_1(\theta) = \left(\frac{\cos kD}{D}\right) - U_1(\theta + 2\pi),\tag{5.33}$$

it is readily seen that the conditions required for the desired solution to the wave equation are satisfied, providing that the expression in Eq. (5.32) approaches zero as the point P approaches infinity; i.e., $D(P, P')$ increases without bound.

The behavior of the correction term in Eq. (5.32) as D increases without bound can be determined by the use of Green's first identity,[5] applied to a large sphere in leaf 2. The branch circle is to be excluded from the sphere by a torus of an arbitrarily small cross-sectional diameter surrounding the circle. Since W_1 is finite on the branch circle, the contribution to an integral from the excluded region is negligible for a sufficiently small diameter.

The relevant Green's identity is

$$\int d\mathbf{r}'' \left[\frac{\sin kD}{D}\Delta'' W_1 - W_1 \Delta'' \frac{\sin kD}{D}\right]$$

$$= \int d\mathbf{r}'' \left[\frac{\sin kD}{D}\left(\Delta'' + k^2\right)W_1 - W_1\left(\Delta'' + k^2\right)\frac{\sin kD}{D}\right]$$

$$= \int dS\left[\frac{\sin kD}{D}\cdot\frac{\partial W_1}{\partial D} - W_1\frac{\partial}{\partial D}\cdot\frac{\sin kD}{D}\right].\tag{5.34}$$

Since $(\sin kD)/D$ is a solution to the wave propagation equation and has no singularity at P', the latter identity may be written as

$$\int d\mathbf{r}''\left(\frac{\sin kD}{D}\right)\left(\Delta'' + k^2\right)W_1(\theta + 2\pi)$$

$$= \int dS\left[\frac{\sin kD}{D}\cdot\frac{\partial W_1}{\partial D} - W_1\frac{\partial}{\partial D}\cdot\frac{\sin kD}{D}\right].\tag{5.35}$$

Since W_1 decreases as $1/D$ for large D while dS varies as D^2, the first term in the surface integral varies as $1/D$ for large D and the integral tends to zero for this term. The second term is equal to $W_1[k(\cos kD)/D] - W_1(\sin kD)/D^2$, and it is clear that the contribution of the second term in the latter difference also tends to zero. However, there is no certainty that the first term in the difference, $dSW_1 k(\cos kD)/D$, tends to zero as presently formulated. However, the latitude in the choice of W_1 can be used to compel the convergence of the latter term to zero.

To this end, the original choice of the integrand of the contour integral constructed in the Sommerfeld method is multiplied by a factor of g, where

$$g = \exp-\left(\frac{kD}{4}\right)\left[\sin\left(u-\theta'\right)+\sin\left(u*-\theta\right)\right]^{2}. \tag{5.36}$$

The factor g serves several purposes: in addition to ensuring convergence as D increases without bound, it also provides convergence on every part of the contour C. Even if the original solution to be generalized is chosen to be $(\exp ikD)/D$ instead of $(\cos kD)/D$, the entire Sommerfeld procedure can be carried out as it was above. With the correction afforded by the volume integral in Eq. (5.32), one then obtains a multiple-valued solution of the wave propagation equation U_1.

If $k = 0$, in both cases U_1 reduces to κ, the unique multiple-valued potential function, which is a multiple-valued generalization of $1/D$. For large values of $D(O,P')$, the product $D(O,P')U_1$ provides an approximation of a multiple-valued generalization of the plane wave $\exp i\mathbf{k}\cdot(\mathbf{r}-\mathbf{r}')$, provided the original solution selected for generalization was $(\exp ikD)/D$.

Finally, it appears that the above discussion can be extended to spaces bounded by branch curves other than the circle, as well as to spaces bounded by more than one branch curve. However, the question of whether the derived solution satisfies the uniqueness conditions of Sec. 3.6 remains open. Since this aspect of the proposed method of solution would lead readers too far from the path of development of Sommerfeld's method as it was originally conceived, it will not be pursued further.

5.4 Diffraction of a Spherical Wave by a Perfectly Conducting Disk

The purpose of solving the boundary-value problem is to determine the electric and magnetic fields corresponding to a spherical wave incident on a perfectly conducting circular disk. The method of solution is essentially the same as that introduced in Chapter 4, though modified somewhat to suit the new configuration.

Thus, Eqs. (4.24) and (4.25) are written as

$$v = U_1\left(\theta\right)-U_1\left(2\pi-\theta\right) \tag{5.37}$$

and

$$W = U_1(\theta) + U_1(2\pi - \theta). \tag{5.38}$$

It is noted that $U_1(2\pi - \theta) = U_1(-\theta)$.

Now, it is defined that

$$\mathbf{M}_\psi = \nabla \times \left(\hat{\mathbf{a}}_z \psi \right), \tag{5.39}$$

where $\hat{\mathbf{a}}_z$ denotes a unit vector along the z-axis with the same sense as the positive z-axis, and the function ψ denotes a solution of the wave propagation equation. In addition,

$$k\mathbf{N}_\psi = \nabla \times \mathbf{M}_\psi, \tag{5.40}$$

$$\mathbf{E} = \mathbf{M}_v + i\mathbf{N}_w, \tag{5.41}$$

and

$$\mathbf{H} = -\mathbf{M}_w + i\mathbf{N}_v. \tag{5.42}$$

It will then be shown that the tangential component of $\mathbf{E}$ and the normal component of $\mathbf{H}$ vanish on the disk $(\theta = +\pi)$.

Calculating $\mathbf{E}$ and $\mathbf{H}$ in terms of v and w, it is found that

$$\mathbf{E} = \hat{\mathbf{a}}_x \left(\frac{\partial v}{\partial y} - \frac{i}{k} \cdot \frac{\partial^2 w}{\partial z \partial x} \right) + \hat{\mathbf{a}}_y \left(-\frac{\partial v}{\partial x} + \frac{i}{k} \cdot \frac{\partial^2 w}{\partial z \partial y} \right) - \hat{\mathbf{a}}_z \frac{i}{k} \left(\frac{\partial^2 w}{\partial x^2} + \frac{\partial^2 w}{\partial y^2} \right), \tag{5.43}$$

and

$$\mathbf{H} = \hat{\mathbf{a}}_x \left(-\frac{\partial w}{\partial y} - \frac{i}{k} \cdot \frac{\partial^2 v}{\partial z \partial x} \right) + \hat{\mathbf{a}}_y \left(\frac{\partial w}{\partial x} + \frac{i}{k} \cdot \frac{\partial^2 v}{\partial z \partial y} \right) - \hat{\mathbf{a}}_z \frac{i}{k} \left(\frac{\partial^2 v}{\partial x^2} + \frac{\partial^2 v}{\partial y^2} \right). \tag{5.44}$$

Since $v = 0$ on the disk $\theta = +\pi$, its derivatives with respect to x and y are also zero on the disk. Because $(\partial w / \partial z)_{(\theta = +\pi)} = (\partial w / \partial \theta)_{(\theta = +\pi)} = 0$, it follows that $E_x = E_y = 0$ on the disk. A similar argument applies to H_z. Hence, the boundary conditions are satisfied.

5.5 Diffraction by a Perfectly Conducting Spherical Dome

The method of solution used above can also be applied to the diffraction of a spherical wave by a perfectly conducting spherical cap. Such a surface is specified by $\theta = \theta_0 (-\pi < \theta_0 < +\pi)$.

The analysis proceeds analogously to that in Sec. 5.4. As before, a vector is defined as

$$\mathbf{M}_\psi = \nabla \times \left(\hat{\mathbf{a}}_\theta \psi \right), \tag{5.45}$$

where $\hat{\mathbf{a}}_\theta$ is a unit vector orthogonal to surfaces $\theta = \text{constant}$, particularly the surface $\theta = \theta_0$.

It is also defined that

$$k\mathbf{N}_\psi = \nabla \times \mathbf{M}_\psi, \tag{5.46}$$

where ψ is a scalar solution of the wave propagation equation.

The solutions of the wave propagation equation are now defined to be

$$v = U_1(\theta) - U_1(2\theta_0 - \theta) \tag{5.47}$$

and

$$w = U_1(\theta) + U_1(2\theta_0 - \theta), \tag{5.48}$$

with

$$\mathbf{E} = \mathbf{M}_v + i\mathbf{N}_w = \nabla \times \left(\hat{\mathbf{a}}_\theta v \right) + \left(\frac{i}{k} \right) \nabla \times \nabla \times \left(\hat{\mathbf{a}}_\theta w \right) \tag{5.49}$$

and

$$\mathbf{H} = -\mathbf{M}_w + i\mathbf{N}_v = -\nabla \times \left(\hat{\mathbf{a}}_\theta w \right) + \left(\frac{i}{k} \right) \nabla \times \nabla \times \left(\hat{\mathbf{a}}_\theta v \right). \tag{5.50}$$

The boundary conditions are

$$\left(\hat{\mathbf{a}}_\eta \right) \cdot \mathbf{E}_{\theta_0} = 0, \ \hat{\mathbf{a}}_\theta \cdot \mathbf{E}_{\theta_0} = 0 \tag{5.51}$$

and

$$\hat{\mathbf{a}}_\theta \cdot \mathbf{H}_{\theta_0} = 0. \tag{5.52}$$

Upon calculation of the field strengths $\mathbf{E}$ and $\mathbf{H}$ [Eq. (5.1)], we find

$$\begin{aligned} \mathbf{E} = &\frac{\cosh\eta - \cos\theta}{a} \left[-\hat{\mathbf{a}}_\eta \frac{\partial v}{\partial \phi} + \left(\sinh\eta \right) \hat{\mathbf{a}}_\phi \frac{\partial v}{\partial \eta} \right] \\ &+ \frac{i\left(\cosh\eta - \cos\theta \right)}{ak} \left[\hat{\mathbf{a}}_\eta \sinh^2\eta \frac{\partial^2 w}{\partial \eta \partial \theta} - \hat{\mathbf{a}}_\theta \left(\frac{\partial^2 w}{\partial \phi^2} + \frac{\partial}{\partial \eta} \sinh^2\eta \frac{\partial w}{\partial \eta} \right) \right], \end{aligned} \tag{5.53}$$

and

$$\mathbf{H} = -\frac{\cosh\eta - \cos\theta}{a}\left[-\hat{\mathbf{a}}_\eta\frac{\partial w}{\partial\eta} + (\sinh\eta)\hat{\mathbf{a}}_\phi\frac{\partial w}{\partial\eta}\right]$$

$$+\frac{i(\cosh\eta - \cos\theta)}{ak}\left\{\hat{\mathbf{a}}_\eta\sinh^2\eta\frac{\partial^2 v}{\partial\theta\partial\eta} - \hat{\mathbf{a}}_\theta\left[\frac{\partial^2 v}{\partial\phi^2} + \frac{\partial}{\partial\eta}(\sinh^2\eta)\frac{\partial v}{\partial\eta}\right] + \hat{\mathbf{a}}_\phi\sinh\eta\frac{\partial^2 v}{\partial\theta\partial\phi}\right\},$$

$$(5.54)$$

where, by a repetition of the reasoning in the preceding section, it can be seen that the boundary conditions are satisfied.

5.6 Comments on the Foregoing Analysis

At this point, it may be asked which of the solutions in Secs. 5.3 and 5.3.3 is the most desirable form of U_1. This question has not yet been answered. The initial goal of this book—to derive sufficient conditions for the solution of boundary-value problems—has been illustrated by the analysis of this chapter.

References

1. A. Sommerfeld, "Mathematische Theorie der Diffraction," *Math. Ann.*, **47**, pp. 317–374 (1896).
2. G. C. Evans, "Lectures on multiple-valued harmonic functions in space," *University of California Publications in Mathematics*, University of California Press, New Series, **1**, pp. 281–340 (1951).
3. A. Sommerfeld, "Über verzweigte Potentiale im Raum," *London Math. Soc. Proc.*, **28**, pp. 395–429 (1897).
4. F. Alzofon, *Multiple-Valued Functions in Three-Dimensional Space and Sommerfeld's Method*, Lockheed Electronics Co., Houston, TX (1970).
5. J. Jackson, *Classical Electrodynamics, Second Edition*, p. 41, John Wiley & Sons, New York (1975).

Chapter 6
Fresnel Diffraction by a Flat Circular Annulus

Thus far, the method of ensuring that the branch points of the distance function D_1 vary as $u_b = \theta + i\alpha$, an essential feature of the Sommerfeld method, has not differed greatly from the procedure used by Sommerfeld.[1] However, generalizing the method from one branch curve to two or more requires a more readily applied formulation than that used in Ref. [1]. The original procedure made use of transcendental functions in an essential manner and provided little indication of how the method could be extended.

In this chapter, the algebraic formulation discussed briefly in previous chapters is generalized in order to deal with several branch curves. As an example of this general approach, a multiple-valued generalization of a point radiation source, defined on a space bounded by two coplanar and concentric branch circles, is constructed. Then, by a procedure described in the preceding chapters, such a solution of the wave propagation equation can be used to construct a solution of a boundary value problem for a perfectly conducting, flat annular ring corresponding to a spherical wave incident on the ring (Fig. 6.1). A minor alteration in the imaging procedure leads to a solution for Fresnel diffraction by a circular annular aperture in a perfectly conducting plane (Fig. 6.2). The boundary conditions can also be imposed on surfaces other than those illustrated in Figs. 6.1 and 6.2 and spanning the branch circles. The surfaces need only be specified by a constant value $\theta = \theta_0$, as is discussed in Sec. 5.5 and Eq. (6.3).

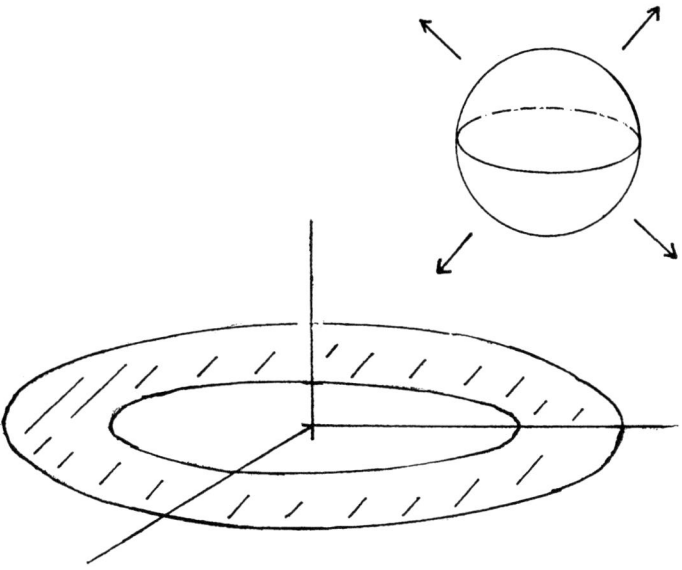

Figure 6.1 Spherical wave incident on a planar perfectly conducting annular ring.

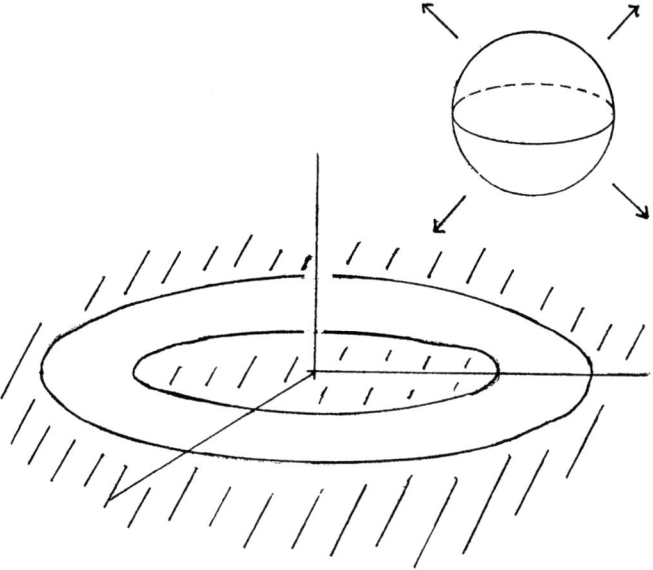

Figure 6.2 Spherical wave incident on a circular annular opening in a perfectly conducting plane.

6.1 Outline of the Generalized Sommerfeld Method

The construction of a suitable coordinate system is central to the success of the Sommerfeld method. For every case previously discussed, the coordinate system was generated by introducing line charges perpendicular to a plane. The equipotential curves in the latter plane, represented by $\eta = \mu = $ constant and the lines of force $\theta = v = $ constant, form a set of mutually perpendicular curves that can be used as a coordinate system. The angle variable can then be used to define a multileaved space.

All the equipotential curves are closed and contain source-line intersections with the plane in their interiors. All the lines of force terminate on the line charges. The functions η and θ are conjugate parts of a complex analytic function.

In the toroidal coordinate system of Chapter 5, for example, describing a space bounded by a branch circle, the function $\zeta = \exp(-\eta + i\theta)$ mapped the region $r > 0$ on the interior $(\eta > 0)$ of a circle with the value $\eta = +\infty \, (r = a)$ corresponding to the circumference of the circle. The value $\eta = 0 \, (r = 0)$ corresponds to the z-axis. Alternatively, one may consider $1/\zeta$ as the mapping function for the region of $r < 0$ into the interior of a unit circle; the value $\eta = -\infty$ corresponds to the center of this circle and $\eta = 0$ corresponds to its circumference. For both circles, the angle variable θ has a simple interpretation, and the extension of the interval of definition of the angle variable from $-\pi < \theta \le +\pi$ to $-\pi < \theta \le 3\pi$ yields an obvious generalization to a two-leaved space. Although not discussed in this work, the extension to an n-leaved space (n = 1, 2, 3, 4...) is immediate.[3]

The function $\zeta = \exp(-\eta + i\theta)$ plays an essential role in the following discussion. It is closely related to a general mapping theorem originated by Riemann: "The interior T of any simply connected region in a plane, whose boundary contains more than one point, can be mapped in a one-to-one conformal manner on the interior of a unit circle."[4]

The mapping function in this monograph is given by ζ, where $\pm\eta$ is the Green's function for the region (note that the Green's function vanishes on the boundary of the region) and this mapping function is uniquely determined.

The next essential feature of the Sommerfeld method is the analytic continuation of the angle coordinate θ' of the fixed source point P'. This analytic continuation alters the distance function D from a real to a complex function D_2, sometimes further modified to a function D_1 (discussed in Chapter 5). The values of u, denoted by u_b, for which D_2 vanishes (i.e., the branch points of D_2 in the u-plane) play a crucial role in the Sommerfeld method. It is essential that u_b take on the form $u_b = \beta + i\alpha$, and $\theta_b'' = \beta$ must have the same periodicity as θ; in all cases considered, $\beta = 0$. The quantity α, which is part of a mapping function $\zeta_{\zeta_b} = \exp(-\alpha + i\theta)$, is a function of the coordinates of P and P'; the function ζ_b maps those points of the u-plane for which $D_2 = 0$ into the interior, S, of several unit circles, corresponding to the several representations of α. The several regions mapped are bounded by the equipotential $\eta = 0$; for example, Chapter 5 contains two such regions corresponding to the two values $\eta_b'' = \pm\alpha$.

The Green's functions $(\cos kD)/D$ and $(\exp ikD)/D$ can then be represented as Cauchy integrals. The contour of integration is a small circle in the u-plane containing the point $u = \theta'$ in its interior. Deformation of the contour and preservation of the single-valued character of the function represented by the integral requires that either branch cut barriers join the branch points, or barriers may pass through the point at infinity. Deformation of the contour demonstrates a dependence of the contour on θ, and definition of a new function U_1 by selection of a part of the contour provides the desired multiple-valued solution.

Next, the above program is made explicit for a two-leaved space bounded by two branch circles.

6.2 The Coordinate System

The coordinate system is generated by four line charges orthogonal to the extended plane $\theta = $ constant. Charges intersect the $r + iz$-plane at the points $\rho = -a(-\text{charge})$, $-b(+\text{charge})$, $+b(-\text{charge})$, and $+a(+\text{charge})$, as illustrated in Fig. 6.3. The potential field of this distribution can be represented by

$$\frac{\rho - a}{\rho + a} \cdot \frac{\rho + b}{\rho - b} = \exp(-\eta + i\theta) = \zeta, \tag{6.1}$$

where $a > b$. It follows from this relation that

$$\sin\theta = \frac{2(a+b)z(r^2 + z^2 + b^2)}{d_1 d_2 d_3 d_4}, \tag{6.2}$$

$$\cos\theta = \frac{\left[(r^2 + z^2 - a^2)(r^2 + z^2 - b^2) + 4abz^2\right]}{d_1 d_2 d_3 d_4}, \tag{6.3}$$

r=iz - plane

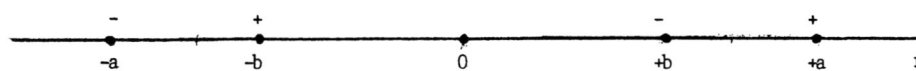

Figure 6.3 Location of line sources generating the coordinate system.

$$\sinh \eta = \frac{2r(a-b)(r^2 + z^2 - ab)}{d_1 d_2 d_3 d_4}, \tag{6.4}$$

and

$$\cosh \eta = \frac{\left[(r^2 + z^2 + a^2)(r^2 + z^2 + b^2) - 4abr^2\right]}{d_1 d_2 d_3 d_4}, \tag{6.5}$$

where $d_1 = |\rho + a|, d_2 = |\rho + b|, d_3 = |\rho - b|$, and $d_4 = |\rho - a|$ [illustrated in Fig. 6.4 with $\theta = (\theta_2 - \theta_1) + (\theta_4 - \theta_3)$].

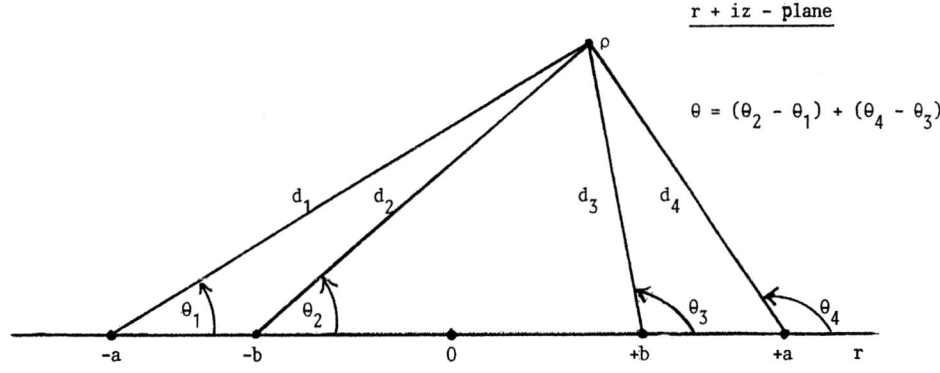

Figure 6.4 Definitions of d₁....d₄ and $\theta_1 .... \theta_4$.

An annular flat surface spans the circles $r = a$ and $r = b$ in the interval $b \leq r \leq a$; the coordinate θ assumes the value $+\pi$ on its upper face and $-\pi$ on its lower face. Since θ is a harmonic function, it follows that it can assume its maximum or minimum value only on the boundaries of the region in which it is defined, i.e., $-\pi < \theta \leq +\pi$.[4]

From Eq. (6.1) it is found that

$$\rho^2 - (a-b)\rho\left[\frac{(1+\zeta)}{(1-\zeta)}\right] - ab = 0 , \qquad (6.6)$$

so that $(\zeta \neq 1)$

$$\frac{2\rho}{a-b} = \frac{1+\zeta}{1-\zeta} \pm \sqrt{\frac{1+\zeta}{1-\zeta} + \frac{2i\sqrt{ab}}{a-b}} \sqrt{\frac{1+\zeta}{1-\zeta} - \frac{2i\sqrt{ab}}{a-b}} . \qquad (6.7)$$

Hence, for given ρ, one value of ζ is determined; for given ζ, two values of ρ are determined. This lack of one-to-one correspondence can be remedied by imagining the values of ζ assigned to a Riemann surface of two leaves (ζ-planes) joined at a branch cut between the points $\zeta_e^\pm$; these values of ζ cause the radicals in Eq. (6.7) to vanish. The corresponding values of ρ, denoted by $\rho_e^\pm = (\pm i\sqrt{ab})$, are also the values for which $(d\zeta/d\rho) = 0$ and are termed *exceptional points*. The equipotential and line of force passing through these points are termed the *exceptional equipotential* and *line of force*.

The exceptional equipotential is composed of two pieces that intersect at a right angle: the circle $|\rho| = +\sqrt{ab}$ and the line $r = 0$ that is specified by $\eta = 0$ (see Fig. 6.5). The exceptional lines of force are specified by

$$\tan\theta_e^\pm = \pm\frac{4(a-b)\sqrt{ab}}{4ab - (a-b)^2} , \qquad (6.8)$$

and each is composed of two pieces lying on the points $\rho_e^\pm$. At each of these points, the exceptional line-of-force pieces lie on the exceptional point, intersect at a right angle, and divide the right angle between the pieces of the exceptional equipotential into equal angles. Except for the exceptional points, the set of couples $\{(\eta, \theta)\}$ together with angle amplitude ψ form a coordinate system in the ρ-plane. The curves $\eta = \text{constant}$ and $\theta = \text{constant}$ are analytic and do not intersect themselves. These curves also intersect one another at right angles except at the exceptional points.

The amplitude angle ψ of the product of the two factors under the radical sign in Eq. (6.7) specifies the leaves of a Riemann surface. The first leaf is specified by the interval $0 \leq \psi < 2\pi$ and corresponds to that part of the ρ-plane for which $r \geq 0$. The second leaf is specified by the interval $2\pi \leq \psi < 4\pi$ and corresponds to $r \leq 0$. The values $\psi = 0$ and 4π are identified (see Fig. 6.5).

The existence of points at which coordinate curves exhibit exceptional behavior is not unique to the proposed coordinate system. The cylindrical coordinate system, for example, specifies the origin of the system by the coordinate $r = 0$ and any of an infinite number of half-lines $\phi = $ constant. In such a case, one may specify the origin uniquely by its coordinates in a rectangular Cartesian coordinate system: $(x, y) = (0, 0)$. A similar device is suggested for the coordinate system $\{(\eta, \theta)\}$ introduced above, if this should prove necessary.

In accord with the Riemann mapping theorem,[4] the function $\zeta = \exp(-\eta + i\theta)$ maps the four regions bounded by the exceptional equipotential $\eta = 0$ into the interiors and exteriors of four unit circles. Subsets of these four regions contain the branch points of the distance function D_2. This is discussed in the next section.

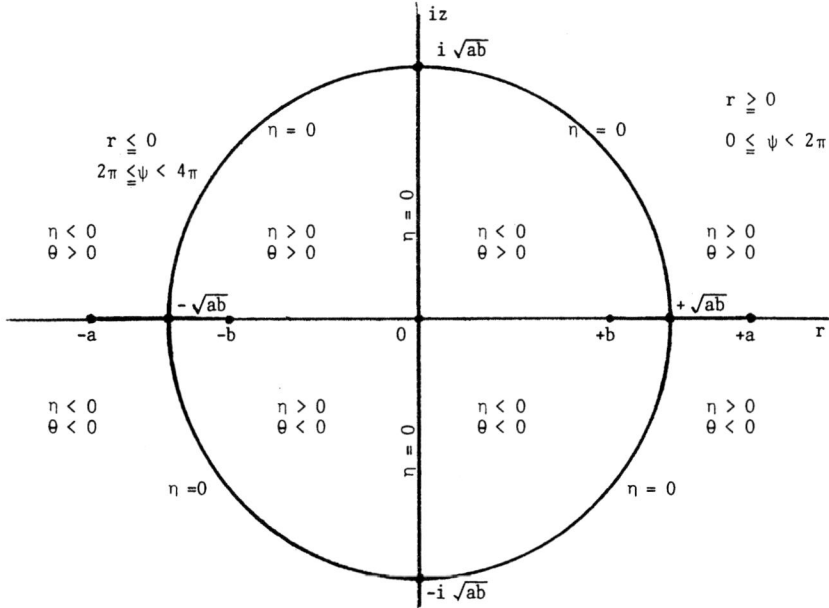

Figure 6.5 Sketch of the exceptional equipotential.

6.3 The Branch Points of D_2

As discussed in Chapter 5, the construction of a multileaved generalization is preceded by an analytic continuation of the coordinate θ' and the determination of the branch points of D_2. A similar procedure, although somewhat generalized, is presented in this section.

The coordinate θ' is continued analytically to the value $u = \theta'' + i\eta''$, inducing an alteration of ρ' to σ, where

$$\frac{\sigma - a}{\sigma + a}\frac{\sigma + b}{\sigma - b} = \exp\left(-\eta' + iu\right) \tag{6.9}$$

$$= \exp\left(-\eta' - \eta'' + i\theta''\right). \tag{6.10}$$

Similarly, $\rho' *$ becomes equal to δ under the same transformation, where

$$\frac{\delta - a}{\delta + a}\frac{\delta + b}{\delta - b} = \exp\left(-\eta' - iu\right) \tag{6.11}$$

$$= \exp\left(-\eta' + \eta'' - i\theta''\right). \tag{6.12}$$

Note that if η'' and θ'' are replaced by $-\eta''$ and $-\theta''$, then σ is replaced by δ. It follows that if σ tends to a, then δ tends to $-a$; similar remarks apply to σ tending to $-a, -b$, and b. Note also that from Eqs. (6.9) and (6.11),

$$\frac{\sigma - a}{\sigma + a} \cdot \frac{\sigma + b}{\sigma - b} = \left[\exp\left(-2\eta'\right)\right]\frac{\delta + a}{\delta - a} \cdot \frac{\delta - b}{\delta + b}, \tag{6.13}$$

providing a relation between σ and δ.

Equations (6.10) and (6.12) map regions of the ρ-plane into the interiors and exteriors of unit circles, as discussed in Sec. 6.2; the coordinate η' has been translated by the addition of η'', and the coordinate θ' has been altered to a variable θ''.

To determine the branch points of D_2, the expression under the radical in

$$D_2 = \sqrt{(\sigma - \rho)(\delta - \rho*) + r(\sigma + \delta)\text{vers}\Phi} \tag{6.14}$$

is set equal to zero, where $\Phi = \phi - \phi'$ and vers $\Phi = 1 - \cos\Phi$. It is then found that with $\sigma = \sigma_b$ and $\delta = \delta_b$,

$$\delta_b = \frac{\sigma_b\left(\rho - r \text{ vers }\Phi\right) - \rho\rho *}{\sigma_b - \rho + r \text{ vers }\Phi}. \tag{6.15}$$

With the aid of Eq. (6.15), one can write

$$D_2^2 = (\sigma - \sigma_b)(\delta - \delta_b) + (\sigma_b - \rho)(\delta - \delta_b) + (\delta_b - \rho*)(\sigma - \sigma_b)$$

$$-\frac{(\sigma - \sigma_b) + (\delta - \delta_b)}{\sigma_b + \delta_b}(\sigma_b - \rho)(\delta_b - \rho*). \quad (6.16)$$

It follows that if $\sigma = \sigma_b$ (implying $\delta = \delta_b$), every term in Eq. (6.16) vanishes separately.

By setting $\sigma = \sigma_b$ and $\delta = \delta_b$ in the expression under the radical sign in Eq. (6.14), it is found that

$$\left[\sigma_b - a - (\rho - a)\right]\left[\delta_b + a - (\rho * + a)\right] + r\left[(\sigma_b - a) + (\delta_b + a)\right] \text{vers } \Phi = 0.$$
$$(6.17)$$

Then, if σ_b tends to a, it can be assumed that ρ also tends to a. Similar considerations apply to σ_b tending to $-a, -b$, and b. More specifically, it can be shown that if σ_b lies in the neighborhood of a, then, approximately,

$$\sigma_b - a \approx (\rho - a)\frac{2\exp(-2\eta')}{\text{vers } \Phi - [\exp(-2\eta')](\cos \Phi + 1)} \quad (6.18)$$

to the first order in $(\rho - a)$. Similar remarks apply to $(\sigma_b + b)$, etc.

It follows that

$$\frac{\sigma_b - a}{\sigma_b + a} \cdot \frac{\sigma_b + b}{\sigma_b - b} = \exp(-\eta' + iu_b) \quad (6.19)$$

$$= \frac{(\rho - a)(\rho + b)}{(\rho + a)(\rho - b)}Q \quad (6.20)$$

$$= [\exp(-\eta + i\theta)]Q, \quad (6.21)$$

where Q is a function that does not vanish or become infinite for finite values of ρ or ρ'; it is an *entire function*.[5]

Setting $u_b = \theta + i\alpha + 2\pi p \, (p = 0, \pm 1, \pm 2...)$, it is found that

$$Q = \exp(\eta - \eta' + \alpha). \quad (6.22)$$

Since Eq. (6.19) maps several regions bounded by the equipotentials $\eta = 0$, there are several expressions for α corresponding to these. Symmetry considerations[3] show that there may be four distinct values of u_b : $-\theta \pm i\alpha_1$ and $\theta \pm i\alpha_2$. The latter values determine the nature of the branch cut barriers that must be introduced in order to employ a Cauchy integral in the u-plane in the construction of a multileaved generalization of a point radiation source. If either α_1 and/or α_2 vanish, the analysis can be modified.

Similar to the case for a single branch circle, there are other branch points, $u_b = \pm i\eta' + 2\pi p \, (p = 0, \pm 1, \pm 2, ...)$, that are independent of θ. The origin of these can be readily understood upon writing

$$\frac{a-b}{2\rho'} = \frac{1-\zeta'}{1+\zeta' \pm \sqrt{1+\zeta' + \dfrac{2i\sqrt{ab}}{a-b}(1-\zeta')} \sqrt{1+\zeta' - \dfrac{2i\sqrt{ab}}{a-b}(1-\zeta')}} \tag{6.23}$$

and noting that, for large σ or δ (i.e., $u_b \approx \pm i\eta'$), $1/D_2$ varies essentially as $1/\sqrt{\sigma}$ or $1/\sqrt{\delta}$. In view of Eq. (6.23), $1/\sqrt{\sigma}$ varies as $[1/\sqrt{2}\sqrt{a-b}]\sqrt{1-\exp(-\eta'+iu)}$ and $1/\sqrt{\delta}$ varies as $[1/\sqrt{2}\sqrt{a-b}]\sqrt{1-\exp(-\eta'-iu)}$, for u approximately equal to u_b. Therefore, to eliminate the latter branch points of $1/D_2$, $1/D_2$ is multiplied by

$$\left[\frac{\sqrt{1-\exp(-\eta'+i\theta')}}{\sqrt{1-\exp(-\eta'+iu)}} \right] \left[\frac{\sqrt{1-\exp(-\eta'-i\theta')}}{\sqrt{1-\exp(-\eta'-iu)}} \right]$$

$$= \frac{\sqrt{\cosh\eta' - \cos\theta'}}{\sqrt{\cosh\eta' - \cos u}} \, .$$

The additional factors ensure that the Cauchy integral with the altered integrand remains equal to $1/D$. Then $D_1 = D_2 \sqrt{\cosh\eta' - \cos u} / \sqrt{\cosh\eta' - \cos\theta'}$ is set.

As discussed in Sec. 5.3, the removal of branch points cannot be done for the factor $\exp ikD_2$ by multiplication by a suitable factor independent of the variable point P. However, adding the term $(\exp -ikD_2)/D_1$ to the term $(\exp ikD_2)/D_1$ and dividing by 2 yields a new integrand: $(\cos kD_2)/D_1$. The value of the Cauchy integral is then $(\cos kD)/D$, and the only remaining branch points are those depending on θ.

Now the branch points $-\theta \pm i\alpha_1, -\theta \pm i\alpha_1 + 2\pi, \theta \pm i\alpha_2$, and $\theta \pm i\alpha_2 + 2\pi$ are joined to the point at infinity by branch cut barriers (Fig. 6.6), where it has been assumed that $\alpha_1 \alpha_2 \neq 0$. There is only one point at infinity in the complex plane. An alternative set of branch cut barriers is illustrated in Fig. 6.7, restricted, however, to the case $\theta \neq \theta'$; the original choice by Sommerfeld of barriers extending to infinity is therefore more convenient. Carrying out the procedure outlined in Sec. 3.3, the desired Green's function U_1 is obtained:

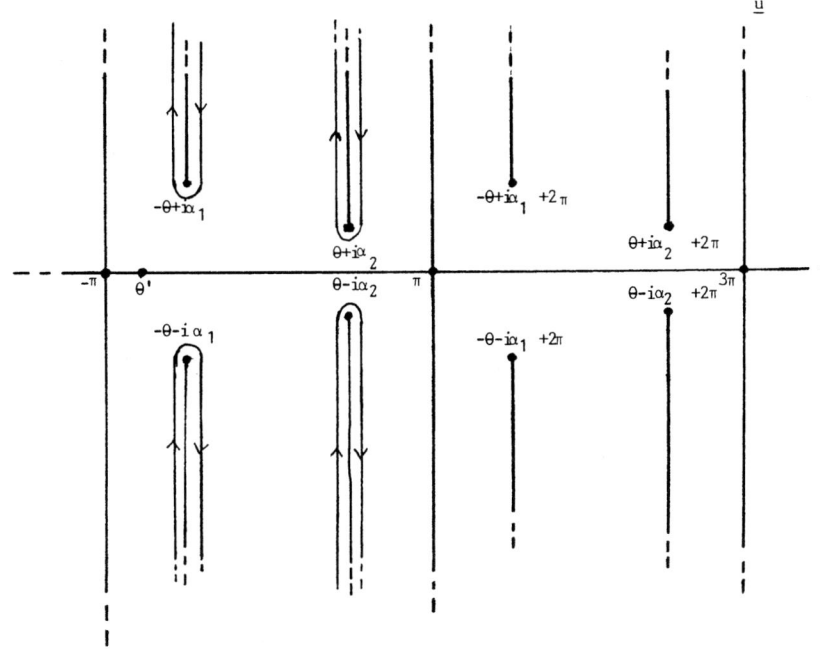

Figure 6.6 Integration contour for branch cut barriers extending to infinity ($\alpha_1\alpha_2 \neq 0$).

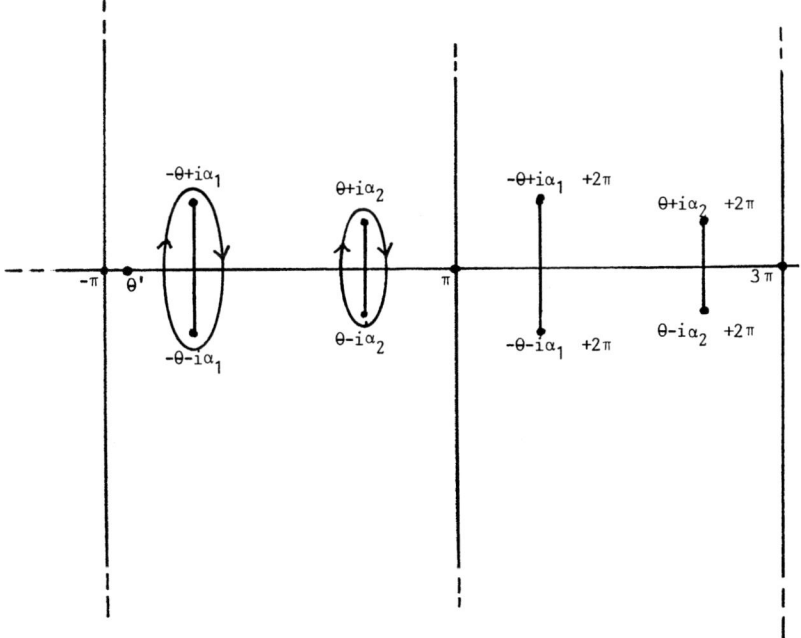

Figure 6.7 Integration contour for finite branch cut barriers ($\theta \neq \theta'$).

$$U_1 = \frac{1}{\pi} \Re \int_{\alpha_1}^{\infty} \frac{i d\eta''}{1 - \exp i(\theta + \theta - i\eta'')/2} \left(\frac{\cos k D_2}{D_1} \right)_{u = -\theta + i\alpha_1 + i\eta''}$$

$$+ \frac{1}{\pi} \Re \int_{d_2}^{\infty} \frac{i d\eta''}{1 - \exp i(\theta' - \theta - i\eta'')/2} \left(\frac{\cos k D_2}{D_1} \right)_{u = \theta + i\alpha_2 + i\eta''} . \qquad (6.24)$$

Note that for large R', with $R' \approx \sqrt{2}(a-b)/\sqrt{\cosh\eta' - \cos\theta'}$, there can be written, where $\eta' \approx 0$ and $\theta' \approx 0$ or 2π,

$$R' U_1(\theta) \approx \cos \mathbf{k} \cdot (\mathbf{r} - \mathbf{r}') - R' U_1(\theta + 2\pi), \qquad (6.25)$$

where $-\pi < \theta \le \pi$ (a multiple-valued generalization of a plane wave). A multiple-valued generalization of $\exp i\mathbf{k} \cdot (\mathbf{r} - \mathbf{r}')$ can be constructed from Eq. (6.25), as in Sec. 5.3.

There appears to be no obstacle to applying the method used above to spaces bounded by more than two branch circles. Moreover, the branch curves need not be circles or straight lines, as will be demonstrated in Chapter 8. Thus, the Sommerfeld method shows considerable promise as a very flexible and general method.

However, as the next chapter will show, if the branch curves extend to infinity, a different kind of analysis is required, although the underlying structure of the Sommerfeld method remains unchanged.

References

1. A. Sommerfeld, "Über verzweigte Potentiale im Raum," *Proceedings of London Math. Soc.*, **28**, pp. 395–429 (1897).

2. M. Born and E. Wolf, *Principles of Optics, Seventh Edition*, Cambridge University Press, Cambridge, UK (1999).

3. F. Alzofon, *Multiple-Valued Functions in Three-Dimensional Space and Sommerfeld's Method*, Lockheed Electronics Co., Houston, TX (1970).

4. O. Kellogg, *Foundations of Potential Theory*, p. 367, Ungar Publication Company, New York (1970).

Chapter 7
Fresnel Diffraction by a Slit between Perfectly Conducting Half-Planes

In this final chapter on the Sommerfeld method, a representation of a radiation source defined on a space bounded by two branch lines is derived. This derivation permits the analysis of Fresnel diffraction by a slit between perfectly conducting half-planes.

The suitability of two coordinate systems that describe spaces bounded by two branch lines is investigated. Only one of these permits the use of Sommerfeld's method.

7.1 Coordinate Systems for Two Branch Lines

The bipolar coordinate system can be used to describe a space bounded by two branch lines.[1] However, despite its simplicity, it is not suitable for the purpose of applying the Sommerfeld method.

The bipolar coordinate system is defined by ($\rho = x + iy$ and $a > 0$):

$$\frac{\rho - a}{\rho + a} = \exp(-\eta + i\theta) = \zeta, \tag{7.1}$$

where $-\infty < \eta < +\infty$ and, to describe a two-leaved space, $-\pi \leq \theta < 3\pi$ (leaf 1 corresponds to $-\pi \leq \theta < +\pi$ and leaf 2 to $+\pi \leq \theta < 3\pi$). A two-sided surface $\theta = \pm\pi$ spans the two branch lines and separates the two leaves. With the replacement of $x + iy$ by $r + iz$, it is clear that the bipolar coordinate system and the toroidal coordinate system of Chapter 5 are formally equivalent. Figures 5.1 through 5.3 apply equally well to Eq. (7.1) with the latter replacement; the toroidal surfaces of Chapter 5 are replaced here by cylinders parallel to the z-axis.

It follows from Eq. (7.1) that

$$\left(\frac{\rho}{a}\right) = \frac{1+\zeta}{1-\zeta}; \tag{7.2}$$

therefore, the distance between two points P and P' is given, in terms of bipolar coordinates, by the positive square root of

$$D^2 = 2a^2 \frac{\cosh\alpha - \cos(\theta - \theta')}{(\cosh\eta - \cos\theta)(\cosh\eta' - \cos\theta)}, \tag{7.3}$$

where $\cosh\alpha = \cosh(\eta - \eta') + \left[(z-z')^2 / 2a^2\right](\cosh\eta - \cos\theta)(\cosh\eta' - \cos\theta')$. Although replacing θ' by $u = \theta'' + i\eta''$ solely in the term $\cos(\theta - \theta')$ of Eq. (7.3) (following the example of Sommerfeld in [Ref. 1]) does yield branch points of the desired kind, i.e., $u_b = \theta \pm i\alpha + 2\pi p \, (p = 0, \pm 1, \pm 2...)$, the function $1/D_3$ obtained in this manner is not harmonic (see Ref. [2] for an example). Additionally, there is no reason to believe that $(\exp ikD_3)/D_3$ is a solution to the equation of wave propagation.

On the other hand, replacing θ' by $u = \theta'' + i\eta''$ everywhere in D would yield a function $(\exp ikD_2)/D_2$, which is a solution to the wave propagation equation but does not have branch points with the desired dependence on θ. This becomes evident upon setting D_2 equal to zero to obtain branch points σ_b and δ_b, where

$$\frac{\sigma_b - a}{\sigma_b + a} = (\exp{-2\eta'}) \frac{(\rho * + a)(\sigma_b - \rho) - (z - z')^2}{(\rho * - a)(\sigma_b - \rho) - (z - z')^2}. \tag{7.4}$$

It can be seen from Eq. (7.4) that if $\rho = a$, it does not follow that $\sigma_b = a$, with a similar observation if $\rho = -a$. It follows that the left side of Eq. (7.4) cannot be proportional to $(\rho - a)/(\rho + a)$ or its reciprocal, and therefore u_b cannot vary as $\pm\theta \pm i\alpha$.

Alternatively, if a new coordinate system is defined by using two positive line charges through the points $\rho = \pm a$ so that $(\theta = \theta_1 + \theta_2)$,

$$(\rho - a)(\rho + a) = a^2 \exp(-\eta + i\theta), \tag{7.5}$$

the relation derived analogously to Eq. (7.4) is

$$\sigma_b^2 - a^2 = (\exp{-2\eta'}) \frac{(\sigma_b - \rho)^2}{\left[\rho * (\sigma_b - \rho) - (z - z')^2\right]^2 - a^2 (\sigma_b - \rho)^2}, \tag{7.6}$$

and it can be shown that if $\rho = \pm a$, then $\sigma_b = \pm a$. Therefore,

$$\sigma_b^2 - a^2 = (\rho^2 - a^2) Q \tag{7.7}$$

$$= a^2 \left[\exp(-\eta + i\theta)\right] Q, \tag{7.8}$$

where Q is a function without zeros in the finite ρ-plane.

With the intervals $-\infty < \eta < +\infty, 0 \le \theta < 4\pi, -\infty < z < +\infty,$ one can define a coordinate system $\{(\eta, \theta, z)\}$ suitable for the analysis of the diffraction of a spherical wave by a slit between two half-planes. The first leaf of the space is specified by $0 \le \theta < 2\pi$, and the second by $2\pi \le \theta < 4\pi$. The half-planes $\theta = 0$ and $\theta = 2\pi$ are depicted in Fig. 7.1

The equipotential curves $\eta = \text{constant}$ are given by

$$\left(x^2 - y^2 - a^2\right)^2 + 4x^2 y^2 = a^4 \exp(-2\eta), \qquad (7.9)$$

and the lines of force $(\theta = \text{constant})$ by the hyperbolas

$$x^2 - y^2 - 2xy \cot\theta = a^2. \qquad (7.10)$$

Except for an exceptional point at $\rho = 0$, where $\partial(-\eta + i\theta)/\partial\rho = 0$, every equipotential curve is analytic and closed, containing one or both of the points $\rho = \pm a$ in its interior. At the exceptional point $\rho = 0$ and in its neighborhood, the exceptional equipotential $\eta = 0$ is divided into two arcs passing through the point with equally spaced tangents, i.e., at right angles to one another.

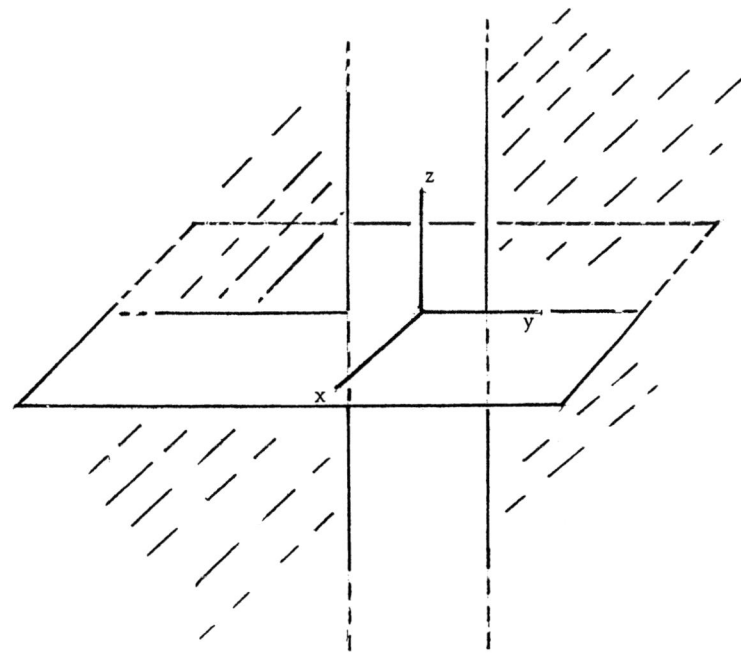

Figure 7.1 Slit between two perfectly conducting half-planes.

Figure 7.2 depicts the ρ-plane and the exceptional equipotential $\eta = 0$, as well as the exceptional line of force $\theta + \pi$. Figure 7.3 presents an approximate sketch of the exceptional equipotential $\eta = 0$ and a more detailed, but approximate, sketch of other equipotentials.

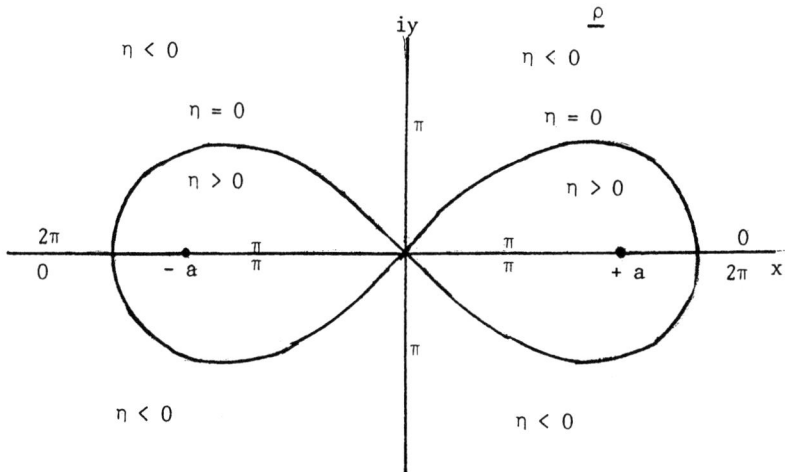

Figure 7.2 The exceptional equipotential and exceptional line of force in the ρ-plane.

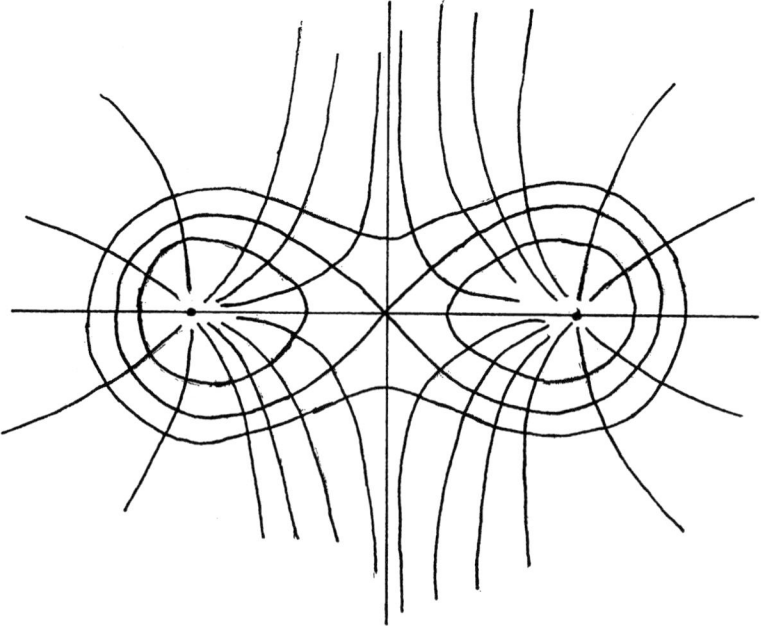

Figure 7.3 Approximate sketch of equipotentials and lines of force.

The ambiguity indicated in the figures regarding the assignment of values for η and θ to regions $x > 0$ and $x < 0$ can be removed by introducing a variable ψ, defined by the equations

$$\rho = a\sqrt{2}\,\exp\frac{-\eta}{2}\left(+\sqrt{\cosh\eta - \cos\theta}\,\right)\exp\frac{i\psi}{2} \qquad (7.11)$$

and

$$\tan\psi = \frac{\left[\exp(-\eta)\right]\sin\theta}{1 + \left[\exp(-\eta)\right]\cos\theta}. \qquad (7.12)$$

If $0 \le \psi < \pi$, then $x > 0$ and $y \ge 0$; if $\pi \le \psi < 2\pi$, then $x \le 0$ and $y > 0$; if $2\pi \le \psi < 3\pi$, then $x < 0$ and $y \le 0$; and if $3\pi \le \psi < 4\pi$, then $x \ge 0$ and $y < 0$.

7.2 Analytic Continuation of θ'

Replacing θ' with $u = \theta'' + i\eta''$, it is found, in accord with Eq. (7.5), that

$$\sigma^2 - a^2 = a^2\,\exp(-\eta' + iu), \qquad (7.13)$$

$$\delta^2 - a^2 = a^2\,\exp(-\eta' - iu), \qquad (7.14)$$

and

$$(\sigma^2 - a^2)(\delta^2 - a^2) = a^4\,\exp(-2\eta'). \qquad (7.15)$$

If $Q = \exp(\eta - \eta' - \alpha)$ is set in Eq. (7.8), it will be found that

$$\sigma_b^2 - a^2 = a^2\,\exp\left(-\eta' - \alpha + i\theta\right); \qquad (7.16)$$

that is, $u_b = \theta + i\alpha + 2\pi p$ $(p = \pm 1, \pm 2...)$. It follows from Eq. (7.15) that

$$\delta_b^2 - a^2 = \frac{a^4\,\exp\left(-2\eta'\right)}{\sigma_b^2 - a^2} \qquad (7.17)$$

$$= a^2\,\exp(\alpha - i\theta - \eta'). \qquad (7.18)$$

From Eq. (7.6) it is seen that there may be four distinct solutions for σ_b, as might have been expected from the nature of the coordinate system. The relation of these roots to one another can be inferred from symmetry considerations.

Thus, given a root σ_{b1}, another root can be derived by noting that replacement of D_2^2, ρ, and $\rho*$ by their conjugates results in

$$(\delta_{b1}^* - \rho)(\sigma_{b1}^* - \rho*) + (z - z')^2 = 0, \qquad (7.19)$$

so that δ_{b1}^{*} is also a solution of the equation $D_{2}^{2} = 0$, i.e., $\delta_{b1}^{*} = \sigma_{b2}$. It follows that

$$\sigma_{b2}^{2} - a^{2} - a^{2} \exp(-\eta' + \alpha_{1} + i\theta);\qquad(7.20)$$

i.e., if $u_{b1} = \theta + i\alpha_{1}$, then $u_{b2} = \theta - i\alpha_{1}$.

Similarly, replacing σ_{b1} by $-\sigma_{b1}$, δ_{b1} by $-\delta_{b1}$, and ρ by $-\rho$, preserves the value of D_{2}^{2}, so that $\sigma_{b3} = -\sigma_{b1}$ and $\delta_{b3} = -\delta_{b1}$. Note that under this transformation the angles ψ and θ increase by 2π. Then the root σ_{b3} varies as $-\rho - a$ nears $\rho = a$ and as $-\rho + a$ nears $\rho = -a$, and is therefore distinct from σ_{b1}. It follows that $(u_{b3} = \theta + i\alpha_{2})$

$$\sigma_{b3}^{2} - a^{2} = a^{2} \exp(-\eta' - \alpha_{2} + i\theta),\qquad(7.21)$$

and by reasoning similar to that above, $u_{b4} = \theta - i\alpha_{2}$, and

$$\sigma_{b4}^{2} - a^{2} = a^{2} \exp(-\eta' + \alpha_{2} + i\theta).\qquad(7.22)$$

This discussion is limited to the case $\alpha_{1}\alpha_{2} \neq 0$.

Corresponding to the choice of roots above, and assuming for the sake of simplicity in illustration that $\alpha_{1} < \alpha_{2}$, barriers are drawn in the u-plane as shown in Fig. 7.4.

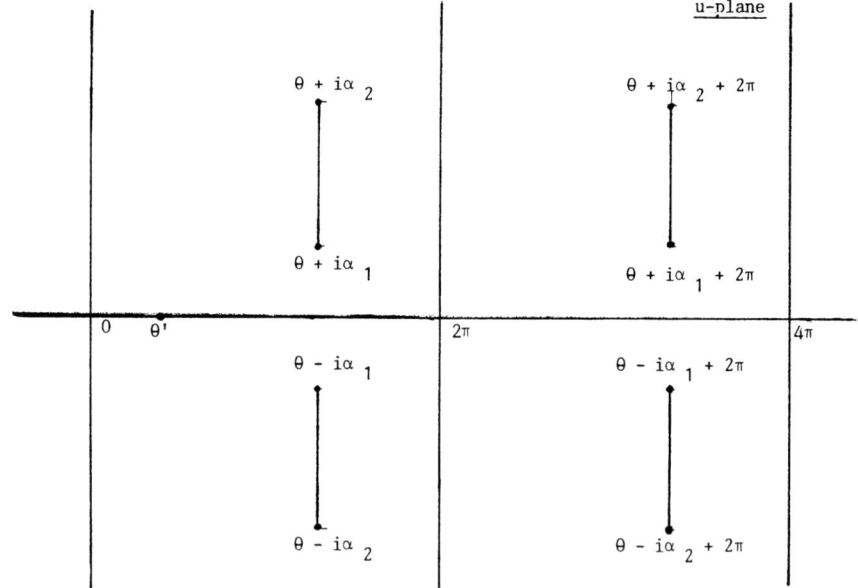

Figure 7.4 Barriers in the u-plane corresponding to branch points $u_{b} = \theta \pm i\alpha_{1}$ and $u_{b} = \theta \pm i\alpha_{2}$ $(\alpha_{1} < \alpha_{2})$.

7.3 Construction of U_1

An essential step in the construction of the two-leaved generalization U_1 of $(\exp ikD)/D$ is the determination of a suitable integrand for the Cauchy integral so that an infinite contour can be introduced. In preceding chapters, it has been shown how this can be done for simpler cases than the one at hand.

To determine what alterations need to be made to the integrand in order to ensure convergence of the resultant integral, consider the behavior of D_2 upon setting $u = \theta'' \pm i\eta_o''(\eta_o'' > 0)$, with η_o'' arbitrarily large. In such a case,

$$\sigma^2 - a^2 \approx a^2 \exp(\mp\eta_o''), \tag{7.23}$$

so that if η_o'' increases without bound (i.e., on the contour to be defined) then for $u = \theta'' + i\eta_o''$, σ^2 tends to a^2 (i.e., σ is real and $\approx \pm a$); while if $u = \theta'' - i\eta_o''$, σ^2 increases without bound (i.e., σ is real and tends to an arbitrarily large positive or negative value). A reference to Eq. (7.15) shows that if σ^2 tends to a^2, then δ^2 increases without bound; and if σ^2 increases without bound, then δ^2 tends to a^2.

In view of the above discussion, it is clear that

$$D_2^2 = (\sigma - \rho)(\delta - \rho*) + (z - z')^2 \tag{7.24}$$

will not provide the desired behavior, as $|u|$ increases without bound, to ensure that $(\exp ikD_2)$—for example, decreases rapidly enough to allow the integrand to force convergence on the integral. However, multiplication of $\exp(ikD_2)$ by $g = \exp{-ka[\cos(u - \theta') - 1]}$ provides an integrand that tends to zero rapidly enough to force convergence, is equal to unity if $u = \theta'$, and does not alter the essential property of the integrand: that it is a solution of the wave propagation equation as a function of the variable point $P(\eta, \theta, z)$.

The procedure described in previous chapters can then be applied to define

$$U_1 = (1/4\pi)\oint \frac{du}{1 - \exp i(\theta' - u)/2} \cdot \frac{g \exp(ikD_2)}{D_2}, \tag{7.25}$$

where C is the contour shown in Fig. 7.5, derived by a process of deforming and deleting as in previous chapters.

The integral in Eq. (7.24) is a solution of the wave equation; it tends to zero as $|\rho|^2 + z^2$ increases without bound; it is single-valued in a two-leaved space bounded by two branch lines, and reduces to a unique multiple-valued potential κ if $k = 0$. In addition,

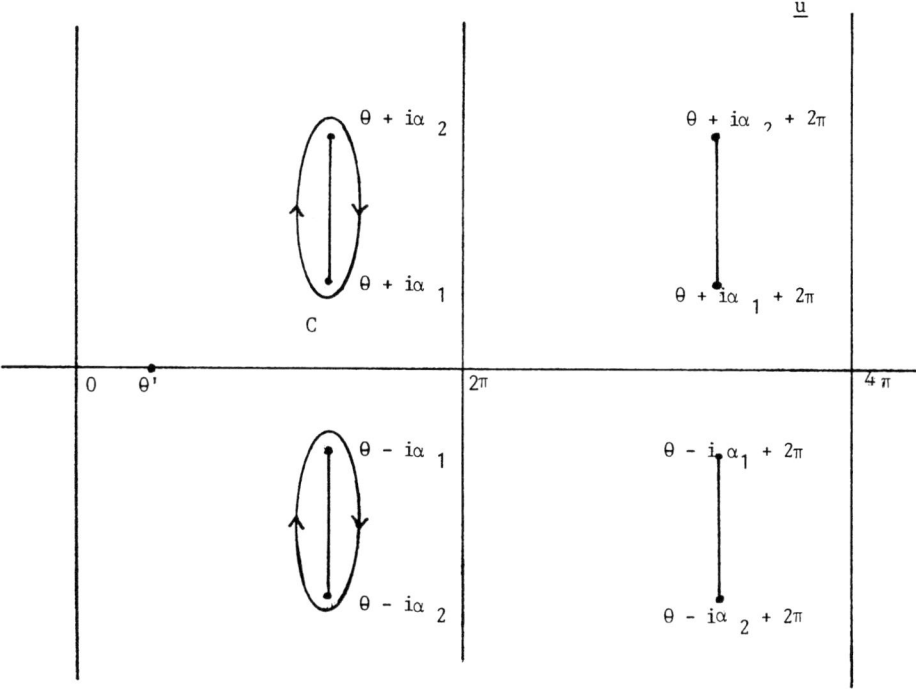

Figure 7.5 The contour C corresponding to branch points $u_b = \theta \pm i\alpha_1$ and $u_b = \theta \pm i\alpha_2$ $(\alpha_1 < \alpha_2)$.

$$U_1(\theta) + U_1(\theta + 2\pi) = \frac{\exp(ikD)}{D}. \tag{7.26}$$

Moreover, for large values of $R' = \sqrt{|\rho'|^2 + z'^2}$ we can derive an approximation for a multiple-valued generalization of a plane wave $\exp i\mathbf{k} \cdot (\mathbf{r} - \mathbf{r}')$, as described in preceding chapters.

7.4 Diffraction of a Spherical Wave by a Slit between Two Perfectly Conducting Half-Planes

The solution of the boundary value problem for a spherical wave incident on a slit between two perfectly conducting half-planes follows the same procedure as that employed in Chapter 4.

Thus,

$$v = U_1(\theta) - U_1(4\pi - \theta) \tag{7.27}$$

and

$$w = U_1(\theta) + U_1(4\pi - \theta), \qquad (7.28)$$

where $U_1(4\pi - \theta) = U_1(-\theta)$, and

$$\mathbf{M}_\psi = \nabla \times (\hat{\mathbf{a}}_y \psi), \qquad (7.29)$$

$$k\mathbf{N}_\psi = \nabla \times \mathbf{M}_\psi, \qquad (7.30)$$

and set

$$\mathbf{E} = \mathbf{M}_v + i\mathbf{N}_w, \qquad (7.31)$$

and

$$\mathbf{H} = -\mathbf{M}_w + i\mathbf{N}_v. \qquad (7.32)$$

The two half-planes are specified by $\theta = 0$. By the same reasoning used in Chapter 4, it can be shown that $E_t = 0$ and $H_y = 0$ on $\theta = 0$, and the boundary conditions are therefore satisfied. Note that $\theta = 2\pi$ does not belong to the first leaf.

Evidently, Fraunhofer diffraction by a slit between two perfectly conducting half-planes can be analyzed by choosing the point source far from the diffracting slit.

7.5 Some Remarks on the Sommerfeld Method

The discussion of Sommerfeld's method to this point has been designed to illustrate the various ways in which the method can be implemented for the configurations of diffracting objects considered. It has been shown that there is an essential difference in the formulation of the solutions of the wave propagation equation for branch curves that extend to infinity and for those that are finite. However, the methods of approach to construction of the multiple-valued Green's functions have been sufficiently general to justify the expectation that the method can be applied to more complex configurations of scatterers.

One great advantage of the Sommerfeld method is its ability to deliver closed analytic expressions for the solution of diffraction problems. In addition, as is indicated in the next chapter, the shapes of the scatterers do not need to be limited to surfaces bounded by straight lines or circles.

It is clear that there is a great deal of valuable research to be done in investigations of optical phenomena with the aid of Sommerfeld's method.

References

1. A. Sommerfeld, "Über verzweigte Potentiale im Raum," *Proceedings of London Math. Soc.*, **28**, pp. 395–429 (1897).

2. F. Alzofon, *Multiple-Valued Functions in Three-Dimensional Space and Sommerfeld's Method*, Lockheed Electronics Co., Houston, TX (1970).

3. A. Sommerfeld, "Mathematische Theorie der Diffraction," *Math. Ann.*, **47**, pp. 317–374 (1896).

4. G. C. Evans, "Lectures on multiple-valued harmonic functions in space," *University of California Publications in Mathematics*, University of California Press, New Series, **1**, pp. 281–340 (1951).

5. M. Born and E. Wolf, *Principles of Optics, Sixth Edition*, Pergamon Press, New York (1980).

6. A. Sommerfeld, *Partial Differential Equations in Physics*, Academic Press, New York (1949).

7. S. Neustadter, "Multiple-valued harmonic functions with circle as branch curve," *University of California Publications in Mathematics*, University of California Press, New Series, **1**, pp. 396–432 (1951).

Chapter 8
Coordinate Systems

In the preceding chapters, the Sommerfeld method has been extended from its original application[1]—a plane wave incident on a perfectly conducting half-plane—to the case of a spherical wave incident on a variety of open surfaces, all perfectly conducting. For sufficiently great distances from the spherical wave source P', the spherical wave approximates a plane wave. Alternatively, in a Fresnel construction, one may imagine a plane wave as the envelope of spherical waves with centers on a plane.

In all cases considered, finding a suitable coordinate system is an essential part of the problem's solution: first, to obtain a branch curve of the desired kind, and second, to obtain branch points that vary correctly with the angle variable θ. To confine oneself to well-known coordinate systems like the cylindrical, bipolar, and toroidal coordinate systems is to severely limit the kinds of branch curves to which this method can be applied, as shown in the foregoing chapters. Moreover, in such a case, the desired goal of providing a flexible tool of analysis for applications in optics is also very limited. In this chapter, it is proposed to show how the Sommerfeld method can be freed of some of the restrictions of coordinate system choice and to extend the discussion to create a new method of analysis that can be applied to closed surfaces of more arbitrary configurations.

8.1 Generalization of the Branch Curves

The coordinate systems discussed so far have led to branch curves with simple shapes: circles and straight lines. However, the Sommerfeld method does not require this simplicity; in this section, it is shown how the branch curves can be generalized.

For example, suppose there is one branch curve: an ellipse in the (x, y)-plane,

$$\frac{x^2}{a^2} + \frac{y^2}{b^2} = 1,$$

(8.1)

with $ab \neq 0$. This can be put into the form

$$r^2 \left[\frac{(\cos^2 \phi)}{a^2} + \frac{(\sin^2 \phi)}{b^2} \right] = 1 \,, \tag{8.2}$$

where r and ϕ are coordinates in a cylindrical coordinate system.

As in Chapter 5, a toroidal coordinate system is now defined with line charges orthogonal to the plane $\phi = \text{constant}$ placed at the points $c(\phi)$ and $c(-\phi)$, where $(0 \leq \phi < \pi)$:

$$c(\pm\phi) = \pm \frac{(ab)^2}{\sqrt{b^2 \cos^2 \phi + a^2 \sin^2 \phi}} \,. \tag{8.3}$$

It is easy to see that

$$r = \frac{c \sinh \eta}{\cosh \eta - \cos \theta} \,, \tag{8.4}$$

$$z = \frac{c \sin \theta}{\cosh \eta - \cos \theta} \,, \tag{8.5}$$

and if $\rho = r + iz$,

$$\frac{\rho - c}{\rho + c} = \exp(-\eta + i\theta) \,. \tag{8.6}$$

As seen in Chapter 5, the relation Eq. (8.6) for given ϕ establishes a one-to-one correspondence between the points of the plane $\phi = \pm\text{constant}$ and the couples $(\eta, \theta) (-\infty < \eta + \infty, \ -\pi < \theta \leq +\pi)$, except for isolated points. Since the space is the sum of these planes, each specified by a unique value of ϕ, it appears, intuitively, that the set of triples $\{(\eta, \theta, \phi)\}$ forms a coordinate system. Alternatively, one may calculate the Jacobian[2] J to verify that since $J \neq 0$, the same conclusion follows. The Jacobian in question is given by

$$J = \begin{vmatrix} \dfrac{\partial r}{\partial \eta} & \dfrac{\partial r}{\partial \theta} & \dfrac{\partial r}{\partial \phi} \\[2mm] \dfrac{\partial z}{\partial \eta} & \dfrac{\partial z}{\partial \theta} & \dfrac{\partial z}{\partial \phi} \\[2mm] \dfrac{\partial \phi}{\partial \eta} & \dfrac{\partial \phi}{\partial \theta} & \dfrac{\partial \phi}{\partial \phi} \end{vmatrix} \,, \tag{8.7}$$

$$J = \begin{vmatrix} \dfrac{\partial r}{\partial \eta} & \dfrac{\partial r}{\partial \theta} \\[2ex] \dfrac{\partial z}{\partial \eta} & \dfrac{\partial z}{\partial \theta} \end{vmatrix} ; \qquad (8.8)$$

and since the form of Eq. (8.8) is the same as would have been obtained in Chapter 5, $Jc^2 = (\cosh \eta - \cos \theta)^{-2}$, it follows that $J \neq 0$.

Although the new coordinate system may not be an orthogonal system, the manner in which the Sommerfeld method is carried out does not depend on the presence of mutually orthogonal coordinate surfaces. The form of the Cauchy integral, for example, is not altered in this case, although it has a different meaning. Indeed, D_2 has the same form as in Chapter 5; however, here it is not an essential requirement.

There appears to be no essential barrier to generalizing the branch curve configuration, so long as the Jacobian does not vanish. For example, one may imagine the method applied to perfectly conducting flakes of very general shape.

8.2 Cylinders of Arbitrary Shape

Although the Sommerfeld method is general in that it is valid for all wavelengths of the incident radiation, it has the fault that it can only be applied to surfaces of negligible thickness; i.e., the wavelength of the incident radiation must be large in comparison with the thickness of the scattering surface.

This section deals with a simplified class of surfaces, i.e., cylinders whose cross sections are bounded by closed curves, to illustrate a method of solution of boundary value problems that does not have the limitations of Sommerfeld's method.

Based on experience, desirable features for a method of solution are (1) an orthogonal coordinate system is used; (2) the scattering body's surface is specified by setting one coordinate equal to a constant; and (3) there is a complete set of elementary solutions of the equations of motion that feature separated variables. That is, any reasonable solution to the boundary-value problem can be expressed as a linear superposition of the members of the set. However, as may be expected, there is a very limited number of coordinate systems for which these solutions can be expected to exist. In the method to be proposed, not all of the features listed can be preserved. Thus, the requirement for an orthogonal coordinate system will be dropped, the second property kept, and a modified form of the third property will be adopted.

For example, suppose that a cylinder is specified by the equation (in a cylindrical coordinate system)

$$r = R_1(\phi) .\tag{8.9}$$

It is assumed that the curve is continuous but need not have continuous derivatives at isolated points on the curve. In order to obtain a coordinate system (i.e., one for which the Jacobian exists and is not zero), the curve with arcs of small circles tangent to the given curve near the isolated points is rounded off. Then the resultant curve has a continuously varying tangent and can be represented by

$$r = R(\phi) ,\tag{8.10}$$

where $R'(\phi)$ is a continuous function. This procedure is used in the following chapter in connection with a boundary-value problem for a hexagonal, solid cylinder.

A family of cylinders is then generated by the equation

$$r = \alpha R(\phi) ,\tag{8.11}$$

where $0 \le \alpha < +\infty$. The surface on which boundary conditions are to be imposed is specified by $\alpha = 1$. The cylindrical coordinate system $\{(r, \phi, z)\}$ is now replaced by the coordinate system $\{(\alpha, \phi, z)\}$. Calculation of the Jacobian yields $J = R$; hence, it is required that $R \ne 0$.

Property (3) requires that a family of elementary solutions exists that forms a complete set; usually the equations of motion are solved in terms of functions selected to yield separable solutions for the given coordinate system. In the present case, this is not done. Instead, elementary solutions suitable for a cylindrical coordinate system are selected, for example, $J_n(kr)(\exp ihz)(\exp im\phi)$, where J_n is a Bessel function of the first kind; these form a complete set of solutions. The choice is motivated by the expectation that the surfaces $\alpha =$ constant and $r =$ constant may approximate one another, although this restriction is not necessary. In the new coordinate system, the elementary solutions have the form $J_n[k\alpha R(\phi)](\exp ihz)(\exp im\phi)$; a detailed example of how these are applied in a boundary-value problem is given in the next chapter.

8.3 Closed Surfaces of Arbitrary Shape

Closed surfaces of arbitrary shape may be approximated by a sphere, for example, although such a restriction is not necessary. In such a case, a new coordinate system is introduced in the same manner as in Sec. 8.2; i.e., a spherical coordinate system $\{(r,\theta,\phi)\}$ is selected ($0 \le r < \infty, 0 \le \theta \le \pi$, $0 \le \phi < 2\pi$), and

$$r = \alpha R(\theta,\phi) \tag{8.12}$$

is set, where $0 \le \alpha < +\infty$ and small sections of spherical surfaces may round off isolated locations of the given surface to yield a continuous, smooth surface $r = R(\theta,\phi)$. Again, it is found that the Jacobian is equal to R and it is required that $R \ne 0$. A complete set of elementary solutions of the wave propagation equation is $j_n[\alpha R(\theta,\phi)]P_{nm}(\cos\theta)\exp im\phi$, where j_n is a spherical Bessel function and P_{nm} is a modified Legendre polynomial.

8.4 Interpolated Coordinate Systems

The method proposed can also be used to construct a coordinate system suitable for imposing boundary conditions on very dissimilar surfaces, e.g., a circular cylinder between two parallel planes. The requirement that the enclosing surfaces be infinite planes can be relaxed, but the simplification is preserved for the purpose of illustration.

The equation of the cylinder is taken to be $r = R(\phi)$, with $0 \le \phi < 2\pi$, $\phi \ne 0$ and $\ne \pi$; the equation of the planes is $\pm y_o = r\sin\phi (y_o > 0)$. One choice for the coordinate system is

$$r = (2 - \alpha)R(\phi) + (\alpha - 1)\left(\frac{y_o}{\sin\phi}\right), \tag{8.13}$$

where $1 \le \alpha \le 2$ specifies the region between the cylinder and the planes, $\alpha = 1$ on the cylinder, and $\alpha = 2$ on the planes. The condition that $\{(\alpha,\phi,z)\}$ be a coordinate system is $y_o \ne R\sin\phi$.

Two complete sets of functions satisfying the equation of wave propagation are $J_n[kr]\exp in\phi \exp ihz$ and $\exp ik_1 x \exp ik_2 y \exp ik_3 z$, with $k^2 = k_1^2 + k_2^2 + k_3^2$ and $x = r\cos\phi, y = r\sin\phi$. Insofar as the formal solution of a boundary value problem is concerned, either set of solutions can be used. However, to calculate conditions close to the cylinder, the first set of solutions is convenient; and for conditions close to the planes, the second set is more convenient. That is, for either set of solutions, a linear superposition of members of the set can be formed, substituting Eq. (8.13) for r in each superposition; boundary conditions are then imposed by setting $\alpha = 1$ on the cylinder and $\alpha = 2$ on the planes. Close to the cylinder, the superposition in terms of Bessel's functions is most convenient for analysis; and for points close to the planes, $\alpha = 2$, the second set of functions is most convenient. However, both superpositions represent the same solution.

In the next two chapters, the above concepts are applied to a specific boundary-value problem with a practical application: diffraction of a plane wave by a hexagonal ice cylinder.

References

1. A. Sommerfeld, "Mathematische Theorie der Diffraction," *Math. Ann.*, **47**, pp. 317–374 (1896).
2. I. Sokolnikoff, *Tensor Analysis*, John Wiley & Sons, New York (1956).

Chapter 9
Radiation Scattering by a Hexagonal Ice Cylinder: Coordinate System

This and the following chapter illustrate the method proposed in the preceding chapter.

The analysis of radiation scattering by a hexagonal water ice cylinder is chosen for its simplicity, as well as for its value to research into a problem of interest in atmospheric phenomena. As well as having a certain simplicity, the boundary-value problem is useful in the study of storms,[1] particularly in the understanding of radiation transfer through ice clouds.

The ability to predict radiation scattering by nonspherical particles is essential in order to provide a complete description of radiation transmission by the atmosphere so that experimental data can be analyzed with certainty. Currently, no general and rigorous method of analyzing the scattering of radiation by solids of given complex shapes is known.[2] For the sake of simplicity in analyzing experimental data, and for the lack of any alternative method of analysis, equivalent Mie scatterers are invoked, i.e., scattering from spherical particles for which an exact analysis is available.

9.1 Configuration

It will be assumed that the ice crystals of interest can be represented in idealized form by infinitely long hexagonal cylinders of ice. In practical terms, this means that the ice crystals met in reality are assumed to be very long in comparison with the wavelength of the radiation incident on them. However, insofar as the analysis to be performed is concerned, there is no restriction on the wavelength of the incident radiation.

The cross section of the hexagonal cylinder, its dimensions, and location relative to the coordinate axes are shown in Fig. 9.1. In order to define a coordinate system, the sharp corners are replaced by small arcs of circles of a radius equal to ε, as shown in Fig. 9.2. Such a procedure has also been recommended by Jones;[3] indeed, it is not likely that a sharp corner would be found in nature.

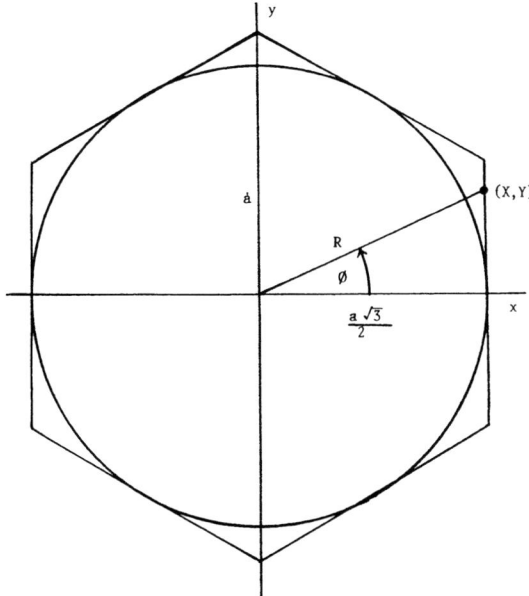

Figure 9.1 Cross section of a hexagonal cylinder and inscribed circle with indicated dimensions.

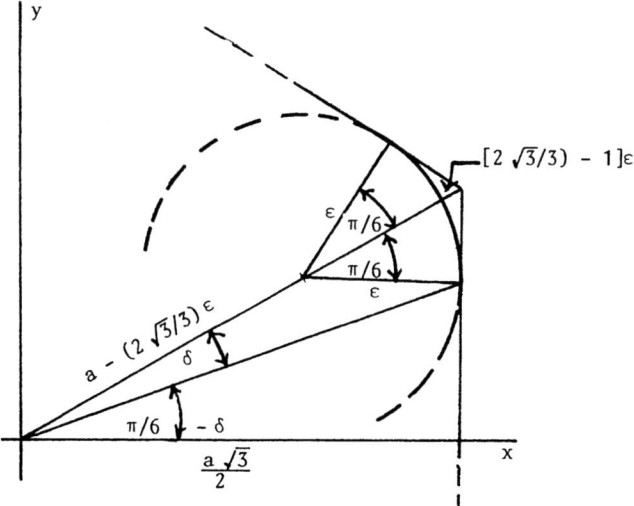

Figure 9.2 Dimensions and angles related to the corner arc of the modified hexagon.

The equation of the modified hexagon is taken to be

$$r = R(\phi), \, 0 \le \phi < 2\pi \tag{9.1}$$

or, alternatively,

$$\left.\begin{array}{l} X = R\cos\phi \\ Y = R\sin\theta \end{array}\right\}; \tag{9.2}$$

and a new coordinate system $\{(\alpha,\phi,z)\}$ is introduced by the relations $(0 \le \alpha < +\infty)$

$$r = \alpha R(\phi),$$

$$\left.\begin{array}{l} x = \alpha X(\phi) = \alpha R\cos\phi \\ y = \alpha Y(\phi) = \alpha R\sin\phi \end{array}\right\}. \tag{9.3}$$

As shown in the preceding chapter, the condition for a valid coordinate system is that $R \ne 0$.

With reference to Fig. 9.1, $R = \left(a\sqrt{3}/2\right)/\cos\phi$ if the angle ϕ lies in the interval $[-(\pi/6)+\delta,(\pi/6)-\delta]$ or for its end points. Similar relations hold if $\phi > \pi/6$. The angle δ specifies the angular extent of the rounded corner; for example, the corner centered on $\phi = \pi/6$ extends from $(\pi/6)-\delta$ to $(\pi/6)+\delta$ (Fig. 9.2).

With reference to the circular arc rounding off the corner at the angle $\phi = \pi/6$, the center of the circle to which it belongs lies at a distance of $a - (2\sqrt{3}/3)\varepsilon$ from the origin of the (x,y)-coordinate system (Fig. 9.2) and a distance $[(2\sqrt{3}/3)-1]\varepsilon$ from the corner of the hexagon. The small circular arc always subtends an angle of $\pi/3$ at the center of the small circle of which it is a part; the length of the arc is therefore equal to $\varepsilon\pi/3$. For small values of ε/a, the angle 2δ subtended by the arc at the origin of the $\{(x,y)\}$ coordinate system is approximately equal to ε/a (Fig. 9.3).

9.2 Unit Vectors

It is convenient to define unit vectors for the new coordinate system analogous to those for a cylindrical coordinate system.

Thus, analogously to the unit vector $\hat{a}_r$ in the sense of increasing r of a cylindrical coordinate system, a unit vector $\hat{a}_\alpha$ is defined in the sense of increasing α of the new coordinate system. Analogously to the angle variable ϕ of a cylindrical coordinate system, an angle-like variable γ of the new coordinate

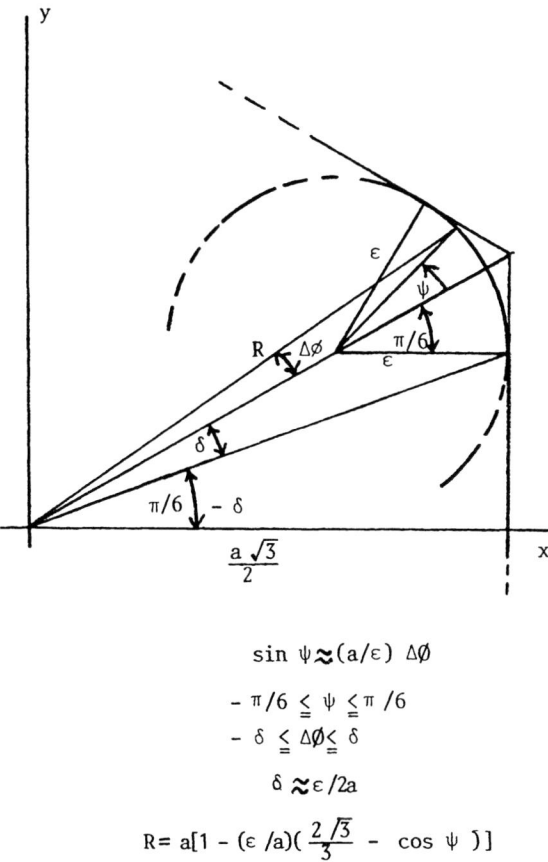

$$\sin \psi \approx (a/\epsilon)\ \Delta\phi$$

$$-\ \pi/6\ \leqq\ \psi\ \leqq \pi\ /6$$

$$-\ \delta\ \leqq\ \Delta\phi \leqq\ \delta$$

$$\delta \approx \epsilon /2a$$

$$R= a[1 - (\epsilon\ /a)(\frac{2\sqrt{3}}{3}\ -\ \cos\ \psi\ )]$$

Figure 9.3 Variation of R on a circular corner of the modified hexagon.

system is defined by

$$\gamma = \left(\frac{1}{L}\right)\int_0^\phi d\phi'\left(\frac{dR}{d\phi'}\right), \qquad (9.4)$$

where

$$L = \int_0^{2\pi} d\phi'\left(\frac{dR}{d\phi'}\right). \qquad (9.5)$$

Corresponding to the angle coordinate γ, a unit vector $\hat{\mathbf{a}}_\gamma$ is defined by

$$\hat{\mathbf{a}}_\gamma = \left(\frac{dX}{d\phi}, \frac{dY}{d\phi}\right)\left(\frac{1}{dR/d\phi}\right). \qquad (9.6)$$

The relation of the unit vectors $\hat{\mathbf{a}}_\alpha$ and $\hat{\mathbf{a}}_\gamma$ to the unit vectors of a cylindrical coordinate system $\hat{\mathbf{a}}_r$ and $\hat{\mathbf{a}}_\theta$ for the interval $-(\pi/6) + \delta \le \phi \le -(\pi/6) - \delta$ is given by

$$\left. \begin{aligned} \hat{\mathbf{a}}_\alpha \cdot \hat{\mathbf{a}}_r &= \cos\phi \\ \hat{\mathbf{a}}_\alpha \cdot \hat{\mathbf{a}}_\phi &= -\sin\phi \\ \hat{\mathbf{a}}_\gamma \cdot \hat{\mathbf{a}}_r &= \sin\phi \\ \hat{\mathbf{a}}_\gamma \cdot \hat{\mathbf{a}}_\phi &= \cos\phi \end{aligned} \right\} . \tag{9.7}$$

In general,

$$\left. \begin{aligned} \hat{\mathbf{a}}_r &= (\hat{\mathbf{a}}_\alpha \cdot \hat{\mathbf{a}}_r)\hat{\mathbf{a}}_\alpha + (\hat{\mathbf{a}}_\gamma \cdot \hat{\mathbf{a}}_r)\hat{\mathbf{a}}_\gamma \\ \hat{\mathbf{a}}_\phi &= (\hat{\mathbf{a}}_\alpha \cdot \hat{\mathbf{a}}_\phi)\hat{\mathbf{a}}_\alpha + (\hat{\mathbf{a}}_\gamma \cdot \hat{\mathbf{a}}_\phi)\hat{\mathbf{a}}_\gamma \end{aligned} \right\} . \tag{9.8}$$

In terms of x and y components, $\hat{\mathbf{a}}_r = (\cos\phi, \sin\phi)$, $\hat{\mathbf{a}}_\phi = (-\sin\phi, \cos\phi)$, $\hat{\mathbf{a}}_\gamma = (0,1)$, and $\hat{\mathbf{a}}_\alpha = (1,0)$ in the interval specified. For other intervals, similar expressions can be derived.

It is not implied that the contribution of the small circular arcs to the scattered electromagnetic field is negligible. Rather, it is expected that the field will vary rapidly near the hexagon's corners. However, some calculations are simplified by neglecting contributions from the circular arcs, as discussed in Chapter 10.

9.3 Inscribed Circle

It is convenient to compare calculations for a scatterer with a hexagonal cross section to those for a scatterer with a circular cross section for which the boundary-value problem has already been solved.[4]

To this end, imagine a circle of radius $a\sqrt{3}/2$ inscribed in the hexagon cross section (i.e., representing a circular cylinder with this radius). Since $R(\phi)$ varies from a minimum of $a\sqrt{3}/2$ to a maximum of about a, $R(\phi)$ and the radius of the circle differ from one another by no more than $[a(1-\sqrt{3}/2)/a] \cdot 100\%$, or about 13% (Fig. 9.1).

For the latter circle, it is evident that $\hat{\mathbf{a}}_r = (\cos\phi, \sin\phi)$ and $\hat{\mathbf{a}}_\phi = (-\sin\phi, \cos\phi)$ for the interval $0 \le \phi < 2\pi$. Evaluation of the differences $\hat{\mathbf{a}}_\alpha - \hat{\mathbf{a}}_r$ and $\hat{\mathbf{a}}_\gamma - \hat{\mathbf{a}}_\phi$ demonstrates that these unit vectors do not differ from one another by a large amount except on the small arcs at the corners of the hexagon.

The above numerical evaluations are useful not only for the bounding surface $\alpha = 1$, but also for description of the radiation field since the radiated field's phase waves are surfaces similar in shape to the surface $\alpha = 1$.

References

1. D. Pitts, "Equivalent blackbody temperature at the top of a severe storm," *J. Appl. Met.*, **14**, pp. 609–618 (1975).
2. A. Holland and G. Gagne, "The scattering of polarized light by polydisperse systems of irregular particles," *Appl. Optics*, **9**, pp. 1113–1121 (1970).
3. D. Jones, *The Theory of Electomagnetism*, Pergamon Press, New York (1964).
4. K.-N. Liou, "Electromagnetic scattering by arbitrarily oriented ice cylinders," *Appl. Optics*, **11**, pp. 667–674 (1972).

Chapter 10
Radiation Scattering by a Hexagonal Ice Cylinder: Boundary Conditions

This chapter discusses the imposition of boundary conditions for a plane wave incident on the modified hexagonal ice cylinder defined in the preceding chapter. Although the resulting formalism is complicated, some simplifications can be made owing to approximations dependent on a restricted radiation wavelength, as well as the small departure of the hexagon from an inscribed circle. These restrictions are not inherent in the method of analysis.

The expressions used in the analysis are modeled on those used in the analysis of radiation scattering by a circular ice cylinder by Liou[1] who asserts that the analysis for radiation scattering by the hexagonal ice cylinder found in nature "appears likely to be impossible to obtain." The intention of this chapter is to show such an analysis is indeed possible.

10.1 Wave Propagation Equation and Elementary Solutions

The wave propagation equation for a homogeneous and isotropic medium is

$$\left(\Delta + m^2 k^2\right)V = 0 , \tag{10.1}$$

where $k = 2\pi / \lambda$ and λ is equal to the wavelength of the radiation in a vacuum, i.e., λ / m is equal to the wavelength of the radiation transmitted through the medium. Here, m is the index of refraction for ice and is a real number for the radiation wavelengths of interest: $0.7\,\mu m$, $3\,\mu m$, and $10\,\mu m$; these are often used in practice. For a wavelength of $0.7\,\mu m$, the index of refraction is equal to 1.31.[1]

The diameters of the ice-cylinder crystals (equal to 2a in Fig. 9.1) encountered in nature are of the same order of magnitude as the incident wavelengths,[1] requiring an accurate description of diffraction effects. The analysis to be presented is valid for all wavelengths in principle, although some approximations are made for greater simplicity.

Elementary solutions of Eq. (10.1) in terms of cylindrical coordinates r, ϕ, and z are ($j^2 = m^2 k^2 - h^2$):

$$\left.\begin{array}{l} J_n(jr)\exp(in\phi)\exp(-ihz) \\ H_n^{(1)}(jr)\exp(in\phi)\exp(-ihz) \\ H_n^{(2)}(jr)\exp(in\phi)\exp(-ihz) \end{array}\right\}, \tag{10.2}$$

where $J_n, H_n^{(1)}$, and $H_n^{(2)}$ are cylindrical Bessel functions ($n = 0, \pm 1, \pm 2, \pm 3 \ldots$), and the functions $H_n^{(1)}$, and $H_n^{(2)}$ are specialized Bessel functions called Hankel functions. A factor of $\exp(i\omega t)$ has been deleted.

Since for the physical parameters given above, the value of ja will be large (here, a is equal to half the diameter of the hexagon discussed in Chapter 9), an approximation of the Bessel functions above can be made by the first term in an asymptotic expansion:

$$J_n(jr) \sim \left(\sqrt{\frac{2}{\pi j r}}\right)\cos\left[jr - \frac{n\pi}{2} - \frac{\pi}{4}\right] \tag{10.3}$$

$$H_n^2(jr) \sim \left(\sqrt{\frac{2}{\pi j r}}\right)\exp\left[-i\left(jr - \frac{n\pi}{2} - \frac{\pi}{4}\right)\right]. \tag{10.4}$$

The set of solutions in Eq. (10.2) is assumed to be complete,[2] i.e., any solution of the wave propagation equation with assigned boundary conditions can be represented by a linear superposition of these elementary solutions. The superposition of functions in Eq. (10.2) is closely related to Fourier-Bessel transforms.[2]

The selection of elementary solutions should conform to physical reality in terms of familiar concepts. In this case, the asymptotic forms of Eqs. (10.3) and (10.4), combined with the remaining factors shown in Eq. (10.2), provide suitable representations of traveling scalar phase waves consistent with the boundary conditions.

The asymptotic representations used below are justified by the following assumptions. Since $\lambda = 0.7\,\mu m$, $k = 8.98\,\mu m^{-1}$; the quantity k has the meaning of the magnitude of a propagation vector of a plane wave incident on the crystal, with components $(h, \ell) = (k\sin\beta, k\cos\beta)$, $0 \leq \phi < \pi/2$. The angle β is chosen equal to 5 deg (.0873 rad) so that $h = 0.782\,\mu m^{-1}$ and $\ell = 8.94\,\mu m^{-1}$, whence it follows that $j = 11.7\,\mu m^{-1}$. Since a is of the order of $5\,\mu m$, $\ell a = 44.7$, and $ja = 58.7$, thereby justifying the use of the asymptotic approximations.

As a consequence of the asymptotic approximation for the Hankel function $H_n^{(2)}$, the asymptotic expression multiplied by the exponential time factor can be interpreted as a cylindrical wave expanding outward from the cylinder $\alpha = 1$ (see Chapter 9) with a phase $\alpha R(\phi) \pm \omega t$ on the surfaces $\alpha R(\phi) \pm \omega t = $ constant. All surfaces for $\alpha = $ constant are similar in shape to one another, and differ only by a scale factor. They are analogous to circular cylindrical phase waves.[1]

10.2 Boundary Conditions

Insofar as the incident plane wave is concerned, the boundary conditions can be divided into two independent kinds: the TM mode corresponds to an incident electric field vector $\mathbf{E}^i$ parallel to the (x, z)-plane, and the TE mode corresponds to an electric field vector orthogonal to the (x, z)-plane. Any other orientation can be expressed as a superposition of these two modes.

For the remaining boundary conditions, the electric and magnetic field vectors can be defined in a manner similar to Eqs. (7.29) through (7.32):

$$\mathbf{M}_\psi = \nabla \times (\hat{\mathbf{a}}_z \psi),\qquad (10.5)$$

$$mk\mathbf{N}_\psi = \nabla \times \mathbf{M}_\psi,\qquad (10.6)$$

$$\mathbf{E} = \mathbf{M}_v + i\mathbf{N}_u,\qquad (10.7)$$

and

$$\mathbf{H} = m(-\mathbf{M}_u + i\mathbf{N}_v),\qquad (10.8)$$

where u and v are independent, suitably chosen solutions of Eq. (10.1).

For the TM mode ($v = v^i = 0$ and $u = u^i$), it is found that

$$\mathbf{E}^i = i\mathbf{N}_u = i\left[\frac{\hat{\mathbf{a}}_x \partial^2 u^i}{\partial x \partial z} + \frac{\hat{\mathbf{a}}_y \partial^2 u^i}{\partial y \partial z} + \hat{\mathbf{a}}_z\left(\frac{\partial^2}{\partial z^2} + m^2 k^2\right)u^i\right];\qquad (10.9)$$

and, in order to ensure that $\mathbf{E}^i$ is parallel to the (x, z)-plane, a scalar wave u^i independent of y is chosen:

$$u^i = (\exp i\omega t)\left[\exp ik(x\cos\beta + z\sin\beta)\right].\qquad (10.10)$$

Similar considerations apply to the TE mode. Diffraction for the TM mode is analyzed below.

A second boundary condition is that the components of electric and magnetic fields tangent to the boundary $\alpha = 1$ must be continuous across the boundary.[1] In terms of the notation of Chapter 9, this condition is written as

$$E_\gamma^i + E_\gamma^s = E_\gamma^t,$$ [10.11(a)]

$$E_z^i + E_z^s = E_z^t,$$ [10.11(b)]

$$H_\gamma^i + H_\gamma^s = H_\gamma^t,$$ [10.11(c)]

and

$$H_z^i + H_z^s = H_z^t.$$ [10.11(d)]

Here superscripts refer to incident (i), scattered (s), and transmitted (t) fields.

The tangential components of Eqs. [10.11(a)–(d)] can be computed from expressions for electric and magnetic field vectors in terms of cylindrical coordinate components along the $\hat{\mathbf{a}}_r, \hat{\mathbf{a}}_\phi$, and $\hat{\mathbf{a}}_z$ unit vectors. Thus, dropping the common factor $\exp(-ihz)$ in any superposition of elementary solutions for Eqs. (10.3) and (10.4), it can be written that[1]

$$\mathbf{E} = \hat{\mathbf{a}}_r \left[\left(\frac{1}{r} \right)\left(\frac{\partial v}{\partial \phi} \right) + \left(\frac{h}{mk} \right)\left(\frac{\partial u}{\partial r} \right) \right] + \hat{\mathbf{a}}_\phi \left[-\left(\frac{\partial v}{\partial r} \right) + \left(\frac{h}{mkr} \right)\left(\frac{\partial u}{\partial \phi} \right) \right]$$
$$+ \hat{\mathbf{a}}_z \left(\frac{i}{mk} \right)\left(m^2 k^2 - h^2 \right) u \tag{10.12}$$

and

$$\mathbf{H} = \hat{\mathbf{a}}_r \left[\left(\frac{h}{k} \right)\left(\frac{\partial v}{\partial r} \right) + \left(\frac{m}{r} \right)\left(\frac{\partial u}{\partial \phi} \right) \right] + \hat{\mathbf{a}}_\phi \left[m\left(\frac{\partial u}{\partial r} \right) + \left(\frac{h}{kr} \right)\left(\frac{\partial u}{\partial \phi} \right) \right]$$
$$+ \hat{\mathbf{a}}_z \left(\frac{i}{k} \right)\left(m^2 k^2 - h^2 \right) v \tag{10.13}$$

(the latter equations differ somewhat from those given in Ref. [1]). It follows that, for example,

$$E_\gamma = \left(\hat{\mathbf{a}}_\gamma \cdot \hat{\mathbf{a}}_r \right) E_r + \left(\hat{\mathbf{a}}_\gamma \cdot \hat{\mathbf{a}}_\phi \right) E_\phi \tag{10.14}$$

with a similar equation for H_γ.

Note that $(n = 0, \pm 1, \pm 2, \pm 3 ...)$, with R given by Eq. (9.1),

$$u^i = \sum_n G_n (\exp in\phi) J_n (\ell\alpha R), \tag{10.15}$$

where $G_n = (-1)^n \exp[i(\omega t - hz)]$, and for $\alpha \geq 1$,

$$u^s = -\sum_n b_n G_n H_n^{(2)} (\ell\alpha R) \exp(in\phi); \tag{10.16}$$

$$v^s = -\sum_n a_n G_n H_n^{(2)} (\ell\alpha R) \exp(in\phi), \tag{10.17}$$

and for $\alpha \leq 1$,

$$u^t = \sum_n d_n G_n J_n (j\alpha R) \exp(in\phi), \tag{10.18}$$

and

$$v^t = \sum_n c_n G_n J_n (j\alpha R) \exp(in\phi). \tag{10.19}$$

All of these expressions for the independent solutions u and v are modeled on Ref. [1] with the exception of the dependence on ϕ.

In connection with solutions of the form assumed above, it is also assumed that the series can be differentiated term by term to yield derivatives of u and v. Here it is noted that

$$\frac{\partial J_n (j\alpha R)}{\partial \alpha} = J_n' (j\alpha R) jR, \tag{10.20}$$

where the notation $J_n'(j\alpha R)$ has the meaning $dJ_n(x)/dx$, with the argument x replaced by $j\alpha R$ after differentiation. In addition,

$$\frac{\partial}{\partial \phi} J_n (j\alpha R) \exp(in\phi) = \left[J_n' (j\alpha R) j\alpha \left(\frac{\partial R}{\partial \phi} \right) + J_n (j\alpha R) in \right] \exp(in\phi). \tag{10.21}$$

10.3 Field Continuity along the z-Axis

The simplest boundary conditions are summarized in Eqs. [10.11(b)] and [10.11(d)]; Eq. [10.11(b)] coupled with Eq. (10.12) implies that

$$\left(\frac{1}{k}\right)(k^2 - h^2)(u^i + u^s) = \left(\frac{1}{mk}\right)(m^2 k^2 - h^2)u^t . \tag{10.22}$$

Upon substituting the sums for u^i, u^s, and u^t from Eqs. (10.15), (10.16), and (10.18), it is found that

$$\left(\frac{1}{k}\right)\left(k^2 - h^2\right)\sum_n G_n\left[J_n\left(\ell R\right) - b_n H_n^{(2)}\left(\ell R\right)\right]\exp\left(in\phi\right)$$

$$\left(\frac{1}{mk}\right)\left(m^2 k^2 - h^2\right)\sum_n G_n d_n J_n\left(jR\right)\exp\left(in\phi\right). \quad (10.23)$$

If the cylinder under consideration were circular, then R would equal a, i.e, the radius of the circular cylinder; and since the series on both sides of the equality in Eq. (10.23) would then be a power series in $\exp(i\phi)$, it follows that the corresponding coefficients of $\exp(in\phi)$ would be equal. However, in the present case, R is a function of ϕ, and the latter simplicity is not possible for the hexagonal cylinder.

Fourier analyzing both sides of Eq. (10.23) and setting, for example,

$$J_n\left(\ell R\right) = \sum_p J_{np}^\ell \exp\left(ip\phi\right), \quad (10.24)$$

where

$$J_{np}^\ell = \left(\frac{1}{2\pi}\right)\int_0^{2\pi} d\phi\, J_n(\ell R)\exp(-ip\phi), \quad (10.25)$$

it is noted that on account of the periodicity of R (with period $\pi/3\,\mathrm{rad}$), that $J_{np}^\ell = 0$ unless $p = 6q\,(q = 0,\pm1,\pm2...)$. In such a case, it is sufficient to set

$$J_{np}^\ell = \left(\frac{3}{\pi}\right)\int_{-\pi/6}^{\pi/6} d\phi\, J_n\left(\ell R\right)\exp\left(-ip\phi\right). \quad (10.26)$$

The latter result is readily generalized to other expressions; it will be assumed that $p = 6q$ below.

Then, Eq. (10.23) becomes

$$\left(\frac{1}{k}\right)\left(k^2 - h^2\right)\sum_p \sum_n G_n\left[J_{np}^\ell - b_n H_{np}^{(2)\ell}\right]\exp i\left(n + p\right)\phi$$

$$= \left(\frac{1}{mk}\right)\left(m^2 k^2 - h^2\right)\sum_n \sum_p G_n d_n J_{np}^j \exp i\left(n + p\right)\phi. \quad (10.27)$$

In order for the coefficients b_n and d_n to be independent of p, it is sufficient that the summation over p be performed first.

Setting $s = n + p\,(s = 0,\pm1,\pm2...)$, the summation over p becomes replaced by a sum over s, a power series in $\exp(is\phi)$. Corresponding terms must then be equal so that $(p = s - n)$

$$\left[\left(\frac{k^2-h^2}{k}\right)\right]\sum_n\left[J_{np}^{\ell}-b_nH_{np}^{(2)\ell}\right]=\left[\frac{m^2k^2-h^2}{mk}\right]\sum_n d_nJ_{np}^j. \qquad (10.28)$$

Similarly, from the boundary condition of Eq. [10.11(d)], it is found that

$$-\left[\frac{k^2-h^2}{k}\right]\sum_n\left[a_nH_{np}^{(2)\ell}\right]=\left[\frac{m^2k^2-h^2}{mk}\right]\sum_n c_nJ_{np}^j. \qquad (10.29)$$

To "solve" Eqs. (10.28) and (10.29), it is sufficient to require that

$$\left[\frac{k^2-h^2}{k}\right]\left[J_{np}-b_nH_{np}^{(2)}\right]=\left[\frac{m^2k^2-h^2}{mk}\right]d_nJ_{np}^j \qquad (10.30)$$

and

$$-\left[\frac{k^2-h^2}{k}\right]\left[a_nH_{np}^{(2)\ell}\right]=\left[\frac{m^2k^2-h^2}{mk}\right]c_nJ_{np}^j. \qquad (10.31)$$

With these conditions, d_n becomes a function of b_n, and c_n becomes a function of a_n. These are to be substituted into the remaining two equations summarizing the boundary conditions in γ, Eqs. [10.11(a)] and [10.11(c)], thus obtaining two linear equations in a_n and b_n, provided similar sufficient conditions are imposed on the latter two equations.

In connection with the sufficient conditions, there is a method of solving linear equations with an infinite number of unknown quantities.[3] It is dependent on an iterative procedure rather than determinants. The success of the method is dependent on the choice of a good first approximation for the unknown quantities; however, the latter method is not used in the present analysis.

In a superposition of functions $G_nH_{np}^{(2)}\exp i(n+p)\phi$, for example, the summation over p is to be carried out before the summation over n.

10.4 The Boundary Conditions on E_γ and H_γ

The boundary conditions expressed in Eqs. [10.11(a)] and [10.11(c)] can be made explicit with the aid of Eqs. (10.12) through (10.14).

It is found that the boundary condition for E_γ becomes

$$\sum_n G_n\left[\exp(in\phi)\right]a_n\left\{(\hat{\mathbf{a}}_\gamma\cdot\hat{\mathbf{a}}_r)\left(-\frac{1}{R}\right)\left[H_n^{(2)\prime}(\ell R)\ell R'+H_n^{(2)}(\ell R)in\right]\right.$$
$$\left.+(\hat{\mathbf{a}}_\gamma\cdot\hat{\mathbf{a}}_\phi)\left[-H_n^{(2)}(\ell R)\ell\right]\right\}$$

$$+ \sum_n G_n \left[\exp(in\phi) \right] b_n \left\{ \left(\hat{\mathbf{a}}_\gamma \cdot \hat{\mathbf{a}}_r \right) \left(-\frac{h}{mk} \right) \left[H_n^{(2)'}(\ell R)\ell \right] \right.$$
$$\left. - \left(\hat{\mathbf{a}}_\gamma \cdot \hat{\mathbf{a}}_\phi \right) \left(\frac{h}{mk} \right) in H_n^{(2)}(\ell R) \right\}$$

$$+ \sum_n G_n \left[\exp(in\phi) \right] c_n \left\{ \left(\hat{\mathbf{a}}_\gamma \cdot \hat{\mathbf{a}}_r \right) \left(-\frac{1}{R} \right) \right.$$
$$\left[in J_n(jR) + J_n'(jR) jR' \right] + \left(\hat{\mathbf{a}}_\gamma \cdot \hat{\mathbf{a}}_\phi \right) J_n'(jR) \Big\}$$

$$+ \sum_n G_n \left[\exp(in\phi) \right] d_n \left\{ \left(\hat{\mathbf{a}}_\gamma \cdot \hat{\mathbf{a}}_r \right) \left(-\frac{h}{mk} \right) J_n'(jR) \right.$$
$$+ \left(\hat{\mathbf{a}}_\gamma \cdot \hat{\mathbf{a}}_\phi \right) \left(-\frac{h}{mk} \right) \left[in J_n(jR) + J_n'(jR) jR' \right] \Big\}$$

$$+ \sum_n G_n \left[\exp(in\phi) \right] \left\{ \left(\frac{h}{mk} \right) \left(\hat{\mathbf{a}}_\gamma \cdot \hat{\mathbf{a}}_r \right) J_n'(\ell R)\ell \right.$$
$$+ \left(\hat{\mathbf{a}}_\gamma \cdot \hat{\mathbf{a}}_\phi \right) \left(\frac{h}{mk} \right) \left[J_n'(\ell R)\ell R + J_n(\ell R) in \right] \Big\} = 0,$$

$$(10.32)$$

and the boundary condition for H_γ becomes

$$\sum_n G_n \left[\exp(in\phi) \right] a_n \left\{ \left(\hat{\mathbf{a}}_\gamma \cdot \hat{\mathbf{a}}_r \right) \left(-\frac{h}{kR} \right) H_n^{(2)}(\ell R)\ell R \right.$$
$$+ \left(\hat{\mathbf{a}}_\gamma \cdot \hat{\mathbf{a}}_\phi \right) \left(-\frac{h}{kR} \right) \left[H_n^{(2)'}(\ell R)\ell R' + H_n^{(2)}(\ell R) in \right] \Big\}$$

$$+ \sum_n G_n \left[\exp(in\phi) \right] b_n \left\{ \left(\hat{\mathbf{a}}_\gamma \cdot \hat{\mathbf{a}}_r \right) \left(\frac{m}{R} \right) \left[H_n^{(2)'}(\ell R)\ell R' + H_n^{(2)}(\ell R) in \right] \right.$$
$$- m \left(\hat{\mathbf{a}}_\gamma \cdot \hat{\mathbf{a}}_\phi \right) H_n^{(2)'}(\ell R)\ell R \Big\}$$

$$- \sum_n G_n \left[\exp(in\phi) \right] c_n \left\{ \left(\hat{\mathbf{a}}_\gamma \cdot \hat{\mathbf{a}}_r \right) \left(\frac{h}{k} \right) J_n(jR) \right.$$
$$+ \left(\hat{\mathbf{a}}_\gamma \cdot \hat{\mathbf{a}}_\phi \right) \left(\frac{h}{kR} \right) \left[in J_n(jR) + J_n'(jR) jR' \right] \Big\}$$

$$- \sum_n G_n \left[\exp(in\phi) \right] d_n \left\{ \left(\hat{\mathbf{a}}_\gamma \cdot \hat{\mathbf{a}}_r \right) \left(-\frac{m}{R} \right) \left[in J_n(jR) + J_n'(jR) jR' \right] \right.$$
$$+ \left(\hat{\mathbf{a}}_\gamma \cdot \hat{\mathbf{a}}_\phi \right) m J_n'(jR) j \Big\}$$

$$+ \sum_n G_n \left[\exp(in\phi) \right] \left\{ (\hat{\mathbf{a}}_\gamma \cdot \hat{\mathbf{a}}_r) \left(-\frac{m}{R} \right) \left[in J_n(\ell R) + J_n'(\ell R) \ell R' \right] \right.$$
$$\left. + (\hat{\mathbf{a}}_\gamma \cdot \hat{\mathbf{a}}_\phi) \left(\frac{m}{R} \right) J_n'(\ell R) \ell R \right\} = 0.$$

(10.33)

With reference to Eqs. (10.32) and (10.33), it is noted that, for example, $J_n'(w) = (1/2)[J_{n-1}(w) - J_{n+1}(w)]$, where w is a real number. A similar relation holds for $H_n^{(2)'}(w)$. If the asymptotic approximations of Eqs. (10.3) and (10.4) are substituted in the example equation for the derivatives, it is found that

$$J_n'(\ell r) \sim \left(\sqrt{\frac{2}{\pi \ell r}} \right) \sin \left(\frac{\ell r - n\pi}{2} \right)$$

(10.34)

and

$$H_n^{(2)'}(\ell r) \sim \left(\sqrt{\frac{2}{\pi \ell r}} \right) \exp \left[-i \left(\frac{\ell r - n\pi}{2} \right) \right],$$

(10.35)

as asymptotic approximations for the derivatives.

Like the procedure followed in Sec. 10.3, each term in Eqs. (10.32) and (10.33) is Fourier analyzed in ϕ, $n + p$ is set equal to s in order to obtain a power series in $\exp(i\phi)$, and the coefficient of each term (an infinite series) set equal to zero, with the understanding that the values of c_n and d_n have the values determined in Sec. 10.3 in terms of a_n and b_n. However, it is worthwhile to evaluate the Fourier transforms of the above terms explicitly; in particular, it is important to estimate the relative contributions of the straight-line segments of the modified hexagon and the circular arcs at the corners.

10.5 Simplifications by Use of Symmetry

In accord with considerations similar to those leading to Eq. (10.26), all the Fourier transform integrals met in the boundary conditions can be written in the general form ($p = 6q, q = 0, \pm 1, \pm 2 ...$)

$$F_{np} = \left(\frac{3}{\pi} \right) \int_{-\pi/6}^{+\pi/6} d\phi F_n(\phi) \exp(-ip\phi),$$

(10.36)

where the integral is extended over a straight-line segment from $\phi = -(\pi/6) + \delta$ to $\phi = (\pi/6) - \delta$ and over circular arcs from $\phi = -(\pi/6)$ to $\phi = -(\pi/6) + \delta$, as well as from $\phi = (\pi/6) - \delta$ to $\phi = (\pi/6)$.

In general, $F_n(\phi) = F_n(-\phi)$, implying that

$$F_{np} = \left(\frac{6}{\pi}\right) \int_0^{\pi/6} d\phi F_n(\phi) \cos p\phi. \tag{10.37}$$

The integral over the straight-line segment from $\phi = -(\pi/6) + \delta$ to $\phi = (\pi/6) - \delta$ can be approximated with an error to the order of ε/a if it is assumed that $R = (a\sqrt{3}/2)/\cos\phi$ over the entire interval $\phi = -(\pi/6)$ to $\phi = (\pi/6)$ and the integral is extended over this interval.

The rest of the integral, over the circular arcs, can be simplified to

$$\left(\frac{6}{\pi}\right) \int_0^\delta d\phi F_n\left[\left(\frac{\pi}{6}\right) - \phi\right] \cos p\left[\left(\frac{\pi}{6}\right) - \phi\right], \tag{10.38}$$

or, alternatively,

$$\left(\frac{6}{\pi}\right) \int_{\pi/6-\delta}^{\pi/6} d\phi' F_n(\phi') \cos p\phi'; \tag{10.39}$$

and if ε is small enough, this can be approximated by

$$\delta\left(\frac{6}{\pi}\right) \frac{\sin(p\delta/2)}{(p\delta/2)} \cos\left(\frac{p\pi}{6}\right), \tag{10.40}$$

and may be neglected, with an error of the order of ε/a or less.

10.6 Evaluation of the Fourier Transforms

Calculation of the Fourier transforms indicated above has central importance in the solution of the boundary value problem. In order to evaluate these transforms, it is useful to contrast two extreme cases rather than to limit the discussion to the parameters listed above for an ice crystal and the wavelength given.

In the first case, the wavelength of the incident radiation is much larger than $a(\lambda \gg a$ or la and $ja \ll 1)$; in the second case, the incident radiation wavelength is much smaller than $a(\lambda \ll a$ or la and $ja \gg 1)$. These two extreme cases correspond to cases often considered in physical optics. A more general formulation is discussed at the end of the chapter.

For $\lambda \gg a$, the scattered radiation depends weakly on the shape of the scatterer. Owing to this circumstance, the resultant field is considered to be a superposition of a field scattered by a circular cylinder (for which a solution is available[1]), and a second field representing a small perturbation because of the small departure of the hexagonal cylinder from a constant radius.

For $\lambda \ll a$, the scattered radiation depends strongly on the shape of the hexagonal cylinder. A perturbation method is not suitable for this case. Since la and ja are much greater than unity, the Bessel functions J_n and $H_n^{(2)}$ are approximated by the first term of an asymptotic expansion: Eqs. (10.3) and (10.4).

10.6.1 Perturbation method

The perturbation calculation is based on the approximation $\left(-\pi/6 \leq \phi \leq \pi/6\right)$

$$R = \left(\frac{a\sqrt{3}}{2}\right)\frac{1}{\cos\phi} \qquad (10.41)$$

$$\approx \left(\frac{a\sqrt{3}}{2}\right)+\left(\frac{a\sqrt{3}}{4}\right)\phi^2\left[1-\left(\frac{\phi^2}{12}\right)+\cdots\right]; \qquad (10.42)$$

and as $\left(\phi^2/12\right) \leq 0.023$, the approximation is valid to the order of 0.02.

Corresponding to the approximation of R $\left(0 \leq \alpha < +\infty\right)$, one sets

$$\alpha R \approx \alpha\left(\frac{a\sqrt{3}}{2}\right)+\alpha\left(\frac{a\sqrt{3}}{4}\right)\phi^2, \qquad (10.43)$$

where it is noted that $\alpha(a\sqrt{3}/2) = r$, a coordinate of a cylindrical coordinate system $\{(r, \phi, z)\}$.

The Bessel function $J_n[\ell r + \alpha\ell(a\sqrt{3}/4)\phi^2]$ can be expanded in the series[3]

$$J_n\left(\ell r + \alpha\ell\frac{a\sqrt{3}}{4}\phi^2\right) = \sum_{-\infty}^{+\infty} J_{n-p}\left(\ell r\right)J_p\left(\ell\alpha\frac{a\sqrt{3}}{4}\phi^2\right) \qquad (10.44)$$

$$= J_n\left(\ell r\right)J_o\left(\ell\alpha a\frac{\sqrt{3}}{4}\phi^2\right)$$

$$+J_1\left(\ell\alpha a\frac{\sqrt{3}}{4}\phi^2\right)\left[J_{n-1}\left(\ell r\right)-J_{n+1}\left(\ell r\right)\right]+\cdots, \qquad (10.45)$$

where it is noted that the difference in the square bracket on the right side of the equation is equal to $2J_n'(\ell r)$, while the function multiplying this factor, for small values of the argument, is approximately $(1/4)\ell\alpha(a\sqrt{3}/4)\phi^2$. It can then be written that

$$J_n\left(\frac{\ell a\sqrt{3}}{2}+\frac{\ell a\sqrt{3}}{4}\phi^2\right) \approx J_n\left(\frac{\ell a\sqrt{3}}{2}\right)$$

$$+\left(\frac{\ell a\sqrt{3}}{4}\right)\phi^2\left[J_{n-1}\left(\frac{\ell a\sqrt{3}}{2}\right)-J_{n+1}\left(\frac{\ell a\sqrt{3}}{2}\right)\right],$$

$$(10.46)$$

with a similar expansion for $H_n^{(2)}\left[(\ell a\sqrt{3}/2)+(a\ell\sqrt{3}/2)\phi^2\right]$ as a Taylor series.

Consequently, Eqs. (10.15) through (10.17) for $\alpha \geq 1$ are reformulated with

$$u^i = \sum G_n J_n\left(\ell r\right)\exp in\phi,$$

$$(10.47)$$

which represents an incident wave unaffected by the difference in shape of a circular and hexagonal cylinder. This makes the approximations similar to Eq. (10.46):

$$u^s \approx u^{s(o)} - \sum b_n^{(o)} G_n\left(\frac{\alpha\ell a\sqrt{3}}{4}\right)\phi^2\left[H_{n-1}^{(2)}\left(\frac{\alpha\ell a\sqrt{3}}{2}\right)-H_{n+1}^{(2)}\left(\frac{\alpha\ell a\sqrt{3}}{2}\right)\right]\exp in\phi$$

$$-\sum b_n^{(1)} G_n H_n^{(2)}\left(\frac{\alpha\ell a\sqrt{3}}{2}\right)\exp in\phi,$$

$$(10.48)$$

where b_n, for example, is set equal to $b_n^{(o)} + b_n^{(1)}$, where $b_n^{(o)}$ corresponds to scattering by a circular cylinder and $b_n^{(1)}$ represents the effect of a small departure from the circular shape. It is assumed that $b_n^{(1)} \ll b_n^{(o)}$.

An approximation for v^s is obtained by replacing $b_n^{(o)}$ and $b_n^{(1)}$ by $a_n^{(o)}$ and $a_n^{(1)}$, respectively. In a similar manner, if $\alpha \leq 1$, then u^i and v^i are obtained by replacing $H_n^{(2)}$ by J_n, l by j, and by changing the minus sign to a plus sign. Since ja is larger than ℓa, more terms in Eq. (10.42) will be required.

The Fourier transforms in the boundary conditions of Secs. 10.3 and 10.4 are then easily evaluated. In this connection, it is noted that $\hat{\mathbf{a}}_\gamma \cdot \hat{\mathbf{a}}_r \approx \phi$, $\hat{\mathbf{a}}_\gamma \cdot \hat{\mathbf{a}}_\phi \approx 1$ to the first order of ϕ.

10.6.2 Fourier transforms for small radiation wavelength

If the incident radiation wavelength is smaller than the diameter of the hexagonal cylinder, e.g., for the ice cylinder case of Sec. 10.1, determining the Fourier transforms needed for satisfying the boundary conditions requires different approximations than those of Sec. 10.6.1.

The first to be noted are the asymptotic expressions for the Bessel functions of Eqs. (10.3) and (10.4), which is a considerable simplification and an implied limitation to the accuracy of the calculations.

A further simplification occurs upon altering the variable of integration from ϕ to R. In this transformation, it is noted that for $0 \le \phi \le \pi/6$,

$$d\phi = \frac{dR}{R} \frac{a\sqrt{3}/2R}{\sqrt{1-(3a^2/4R^2)}}, \tag{10.49}$$

$$\cos\phi = \frac{3a}{4R} + \frac{1}{2}\sqrt{1-\frac{3a^2}{4R^2}}, \tag{10.50}$$

$$\sin\phi = \left(\sqrt{\frac{3}{2}}\right)\left(\frac{a}{2R}-\sqrt{1-\frac{3a^2}{4R^2}}\right), \tag{10.51}$$

$$\exp(-i6q\phi) = \cos 6q\phi - i\sin 6q\phi, \tag{10.52}$$

and

$$\cos 6q\phi = 2^{6q-1}(\cos\phi)^{6q} - \frac{6q}{1!}2^{6q-3}(\cos\phi)^{6q-2} + \cdots \tag{10.53}$$
$$+ \frac{6q(6q-3)}{2!}2^{6q-5}(\cos\phi)^{6q-4} - \cdots,$$

where the series terminates with the term equal to zero.[4] In addition, if q is a positive integer, then

$$\sin 6q\phi = (-1)^{3q+1}(\cos\phi)\left[2^{6q-1}\sin^{6q-1}\phi\right]$$
$$- \frac{(6q-2)}{1!}2^{6q-3}\sin^{6q-3}\phi$$
$$+ \frac{(6q-3)(6q-4)}{2!}2^{6q-5}\sin^{6q-5}\phi - \cdots, \tag{10.54}$$

and the series terminates with the term equal to zero.[4] Moreover, in the interval $0 \le \phi \le \pi/6$, $\hat{\mathbf{a}}_\gamma \cdot \hat{\mathbf{a}}_r = \sin\phi$, and $\hat{\mathbf{a}}_\gamma \cdot \hat{\mathbf{a}}_\phi = \cos\phi$.

Also note that in the interval $0 \le \phi \le \pi/6$, R varies by no more than $a[1-(\sqrt{3}/2)]$, or about 13% of the radius of the hexagon, a. Both extreme values,

a and $a\sqrt{3}/2$, vary from their average value $R_a = (a/2)[1+\sqrt{3}/2]$ by approximately 6.7% of a. It is therefore suggested that in order to estimate the value of the Fourier transform integral of Eq. (10.37), all terms except those that are functions of ℓR or jR are estimated by replacing R everywhere by R_a. In view of the asymptotic approximations of the Bessel functions J_n and $H_n^{(2)}$, which can be expressed as sums of exponents $\exp\pm i\ell R$ and $\exp\pm ijR$, it becomes possible to evaluate the Fourier transforms explicitly. The resultant expressions are complicated but can be evaluated in numerical terms. This will not be carried out here.

Thus, in principle, the boundary-value problem can be solved numerically, with the approximations listed above.

10.6.3 Trigonometric interpolation

Because of the periodicity of the integrands of the Fourier integrals, the method of trigonometric interpolation is particularly well suited to calculate the integrands. Moreover, the results are derived in the form of a sum of exponentials that makes it easy to provide explicit expressions for the integrals themselves for all values of la and ja.

Trigonometric interpolation provides a least squares fit to the data; "the numerical procedure...is simple and straight forward, and...well convergent."[5] Since the method is detailed in Lanczos,[5] it is not derived here; only some salient results are presented.

Thus, given $2N+1$, equally spaced values of the integrand $F_n \exp(-ip\phi)$, sampled at the angles $\phi_j = j(\pi/6N)$ with $j = 0,\pm1,\pm2...,\pm(N-1),\pm N$, there are $2N$ equal intervals and ϕ_j varies between $-\pi/6$ and $+\pi/6$. The number of samples required for accuracy depends on the form of the integrand.[5] Then $\left(p = 6q', q' = 0,\pm1,\pm2,\pm3...\right)$[5]

$$F_n\left(\phi\right)\exp\left(-ip\phi\right) = \sum_{q=-N}^{+N} c_q \exp\left(+iq\phi\right) \qquad (10.55)$$

and

$$c_q = \left(\frac{1}{2N}\right)\sum_{J=-N}^{j=+N}{}' F_n\left(\phi_j\right)\exp\left(ip\phi_j\right)\exp\left(-iq\phi_j\right), \qquad (10.56)$$

where the primed sum indicates that the first and last terms of the sum are to be multiplied by $1/2$. Since $F_n(\phi) = F_n(-\phi)$, simplification in the derived expressions is possible, but for the present purpose is not necessary. Then,

$$\int_{-\pi/6}^{\pi/6} d\phi F_n(\phi) \exp(-ip\phi) = \sum_{q=-N}^{q=+N} c_q \left(\frac{\pi}{3}\right) \left[\frac{\sin(q\pi/6)}{q\pi/6}\right], \qquad (10.57)$$

approximately.

It follows that it is possible to calculate the Fourier transforms required for solution of the boundary value problem for large, small, and intermediate values of ℓa and ja.

References

1. K.-N. Liou, "Electromagnetic scattering by arbitrarily oriented ice cylinders," *Appl. Optics*, **11**, pp. 667–674 (1972).

2. H. Margenau and G. Murphy, *The Mathematics of Physics and Chemistry*, D. van Nostrand, New York (1943).

3. National Bureau of Standards, Dept. of Commerce, *Handbook of Mathematical Functions*, Applied Mathematics Series, **55**, Ninth Printing (1970).

4. H. Dwight, *Tables of Integrals and Other Mathematical Data, Fourth Edition*, The Macmillan Co., New York (1964).

5. C. Lanczos, *Applied Analysis*, Prentice Hall, Inc., Englewood Cliffs, NJ (1956).

Appendix A
Alternative Methods of Exact Diffraction Analyses

As indicated in the text, the Sommerfeld solution is only one of a number of methods for analyzing diffraction problems that are exact with respect to wavelength. The purpose of this appendix is to place the Sommerfeld method in historical perspective relative to the development of some of these alternatives and to outline a few of them.

A.1 General Comments

All the exact theories and numerical techniques for computing the scattered electromagnetic field, like the Sommerfeld method, are based on solving the Maxwell equations. Most are numerical in nature, but those depending on the separation of variables result in an analytical solution for only a few simple cases. In contrast, Sommerfeld's method delivers an analytic solution for all cases of interest.

A.2 Historical Development of Some Diffraction Solutions: Post-Sommerfeld

This section describes the development of methods other than Sommerfeld's method to solve diffraction problems. It is largely quoted or paraphrased from Ref. [1] and stresses the importance of Sommerfeld's influence on research subsequent to his original work.

According to Ref. [1], the problem of producing an exact and closed-form solution to any diffraction problem was difficult and often impossible. The best-known and most successful example was Sommerfeld's solution. Later, Sommerfeld's solution for a half-plane was extended to an infinite wedge (Ref. [2]).

Following the latter extension, an integral equation formulation of the original Sommerfeld boundary-value problem was presented, although the integral equation was not solved (Ref. [3]). Eventually the latter integral equation was solved (Ref. [4]). Independently, the same problem was formulated as an inhomogeneous Wiener-Hopf integral equation and a solution given (Ref. [5]). The problem has also been formulated in terms of dual-integral equations that can be easily solved (Ref. [6]). It was felt that integral equation methods could more easily be generalized then the Sommerfeld method, although the equations are difficult to solve.

A diffraction problem closely related to Sommerfeld's occurs when the scatterer is an infinite plane slit. The first solution to this problem was presented in Ref. [7], using a method similar to that of Sommerfeld. The same problem was solved in Ref. [8] by separating the wave equation in elliptical coordinates. Separation of variables in oblate spheroidal coordinates was also employed to solve the problem of diffraction by a circular aperture in a plane screen (Refs. [9] and [10]).

A.3 Modern Alternatives to Sommerfeld's Method

The availability of fast computers has made many numerical methods of solution possible. A few of these alternatives are listed below along with a brief description of their characteristics. They are frequently tailored for a specific problem. The information has been abstracted from Ref. [11].

A.3.1 Finite-element method

The finite-element method computes the scattered time-harmonic electric field by solving the vector Helmholtz (or wave propagation) equation, subject to boundary conditions on the scatterer surface.

The scatterer is embedded in a finite computational domain discretized into many small-volume cells called elements—with about 10 to 20 elements per wavelength. The unknown field values are specified at the nodes of these elements. Upon imposing the boundary conditions, the differential equation is converted into a solvable matrix equation.

A.3.2 Integral-equation method

The scattering of a plane electromagnetic wave can be described by an integral equation.

Calculation of the scattered field depends on determination of the field intensity internal to the scatterer. The interior is divided into 10 to 20 cells per wavelength; the field within each cell is assumed to be constant. This results in a matrix equation that can be solved numerically.

Several ways of dealing with the interaction between source elements within the scatterers have given rise to various methods of solution. These are

(1) The discrete dipole approximation, or coupled dipole method;

(2) The digitized Green's function algorithm;

(3) The volume integral equation formulation; and

(4) The variational volume integral method.

A.3.3 The T-matrix

The T-matrix approach is based on expanding the incident field in vector spherical-wave functions that are regular at the origin and the scattered field in the same kind of functions, regular at infinity. This method is best suited to scatterers with axial symmetry.

References

1. J. Miles, "On the diffraction of an electromagnetic wave through a plane screen," *J. Appl. Phys.*, **20**, pp. 760–771 (1949).

2. H. Macdonald, *Electric Waves*, p. 186, Cambridge University Press, London (1902).

3. W. Magnus, "Über die Beugung electromagnetischer Wellen an einer Halbebene," *Zeitshcr. für Phys.*, **117**, pp. 168–179 (1941).

4. E. Copson, "On an integral equation arising in the theory of diffraction," *Quart. J. Math.*, **17**, pp. 19–34 (1946).

5. J. Schwinger, *Seminar on the Theory of Guided Waves*, MIT Radiation Laboratory (1944).

6. M. Born and E. Wolf, *Principles of Optics, Sixth Edition*, pp. 564–578, Pergamon Press, New York (1980).

7. K. Schwarzchild, "Die Beugung und Polarisation des Lichts durch einen Spalt," *Math. Ann.*, **55**, pp. 177–247 (1902).

8. M. Strutt, "Lamesche, Mathieusche und verwandte Funktionen in Physik und Technik," *Ergeb. d. Math.*, **1**, pp. 202 et al. (1932).

9. H. Bateman, *Electrical and Optical Wave Motion*, Cambridge University Press, London (1915).

10. R. Spence and A. Leitner, "Diffraction by circular disks and apertures," *Phys. Rev.*, **74**, p. 349 (1948).

11. M. Mishchenko, J. Hovenier, W. Wiscombe, and L. Travis, "Overview of scattering by nonspherical particles," in *Light Scattering by Nonspherical Particles, Theory, Measurements, and Applications*, M. Mischenko, J. Hovenier, and L. Travis, Eds., pp. 23–60, Academic Press, New York (1999).

Appendix B
Sommerfeld's Original Analyses

This appendix presents an abbreviated account of Sommerfeld's original analyses using multiple-valued solutions of differential equations.

Corresponding to the two differential equations under consideration, Sommerfeld used two methods of approach. The first was published in a paper concerning the diffraction of a plane wave by a perfectly conducting half-plane,[1] and the second was published in a paper dealing with the solution of Laplace's equation defined on spaces bounded by one and two branch lines.[2] Unfortunately, the function derived for two branch lines was not a solution of Laplace's equation: Sommerfeld was incorrect in his analysis, as shall be shown.

Throughout this book, the term "Sommerfeld's method" has been used in reference to both Refs. [1] and [2] for the sake of brevity, although these approaches are very different from one another. However, both papers present alternate methods of constructing multiple-valued solutions of physical problems: one suited to diffraction and the other to static fields. This author has extended the method used for constructing a solution for static fields to diffraction problems, since it is more direct than the procedure adopted in Ref. [1] and it is more adaptable to the generalization for several branch curves, i.e., more complex scattering configurations. This appendix provides the reader with an opportunity to compare the discussion in this monograph with Sommerfeld's original papers, and why it was difficult to generalize the method.

For simplicity, Sommerfeld's second published paper will be discussed first.

B.1 Static Fields

Sommerfeld states that in order to understand the behavior of static fields, it is convenient to focus attention on the Green's function solution of Laplace's equation. In ordinary, or physical, space without branch curves and for an infinite region, this is the inverse distance $1/D(P,P')$, where P is the point of observation of the field and P' is a source point.

To derive a solution (i.e., Green's function) defined on a multileaved space, Sommerfeld chose a space bounded by one branch curve: a straight line. Since an angle is the most convenient numerical means of specifying the leaves of the three-dimensional space, the branch line was chosen as the z-axis of a cylindrical-coordinate system $\{(r,\phi,z)\}$ with $0 < r < +\infty$, $0 \leq \phi < 2\pi$, and $-\infty < z < +\infty$. The branch curve $r = 0$ is excluded from the space: it is the common boundary of all the leaves of the space. Another boundary of the space is the fixed source point P'; it lies in the first leaf of the space that is identified with real or physical space and $D \neq 0$.

In the cylindrical-coordinate system,

$$D^2 = r^2 + r'^2 - 2rr' \cos(\phi' - \phi) + (z - z')^2 \tag{B.1}$$

$$= (2rr')[\cosh\alpha - \cos(\phi' - \phi)], \tag{B.2}$$

where $\cosh\alpha = (1/2)[(r/r') + (r'/r)] + (z - z')^2/2rr'$ and $\alpha \geq 0$.

It was Sommerfeld's inspiration to alter the dependence of $1/D$ on ϕ' by replacing ϕ' with a complex variable u so the resulting function remained a solution of Laplace's equation with enough flexibility in form to construct a function single-valued on a multileaved space.

The added flexibility was gained by expressing $1/D$ as a Cauchy integral (see Appendix C):

$$\frac{1}{D} = \left(\frac{1}{2\pi n}\right)\oint du \frac{1}{1 - \exp i(\phi' - u)/n} \cdot \frac{1}{D_1}, \tag{B.3}$$

where n is a positive integer, $D_1 = \sqrt{rr'}\sqrt{\cosh\alpha - \cos(u - \phi)}$, and the contour of integrations is a small circle with the point $u = \phi'$ in its interior. The contour is now deformed to the same shape as exhibited in Fig. 3.9; then a new function is defined by selecting only one pair of the hairpin contours. The new function is shown to be a Green's function defined and single-valued on an n-leaved space $(n = 1,2,3...)$.[3] An essential part of the proof is that the branch points $u = u_b$ of D_1 are given by $u_b = \phi \pm i\alpha + 2\pi p$ $(p = 0, \pm1, \pm2, \pm3...)$.

Sommerfeld then extended the construction used for one branch line to two branch lines, using a bipolar coordinate system $\{(\eta, \theta, z)\}$, $-\infty < \eta < +\infty$, $-\pi < \theta \leq +\pi$, $-\infty < z < +\infty$ with[2]

$$x = \frac{a\sinh\eta}{\cosh\eta - \cos\theta}$$

$$y = \frac{a\sin\theta}{\cosh\eta - \cos\theta} \Bigg\} . \qquad (B.4)$$

$$z = z$$

The branch lines lie on the points $x = \pm a$ (or $\eta = \pm\infty$), $y = 0$, and $z = 0$ and are parallel to the z-axis.

In the bipolar coordinate system,

$$D^2 = 2a^2 \frac{\cosh\alpha - \cos(\theta' - \theta)}{(\cosh\eta - \cos\theta)(\cosh\eta' - \cos\theta')}, \qquad (B.5)$$

where $\cosh\alpha = \cosh(\eta - \eta') + [(z - z')^2 / 2a^2](\cosh\eta - \cos\theta)(\cosh\eta' - \cos\theta')$. If the same procedure is followed as for one branch line, a complex variable u is substituted for θ' everywhere in $1/D$ with the certainty that the resulting function would be a solution of Laplace's equation as a function of the point P. However, there is no assurance that the branch points of D will have the form $u_b = \theta \pm i\alpha + 2\pi p$, as before. Instead, Sommerfeld substituted u for θ' only in the term $\cos(\theta' - \theta)$ so that u_b would have the correct dependence on θ and α, sacrificing the certainty that $1/D_1$ would remain a solution of Laplace's equation. Indeed, it can be shown that the resulting function is not a solution of Laplace's equation.[3]

Sommerfeld's difficulties in dealing with the transcendental functions appearing in D were apparently encountered by subsequent attempts to generalize and systematize Sommerfeld's initial success for the case of one branch line. However, the method was successful for the case of one branch circle, owing to the formal similarity of bipolar and toroidal coordinate systems and the lack of dependence of $\cosh\alpha$ on the leaf angle. Thus, for toroidal coordinates, D^2 has the same form as in Eq. (B.5), but $\cosh\alpha = \cos(\eta - \eta') + \sinh\eta\sinh\eta'[1 - \cos(\phi - \phi')]$.

In contrast, reformulation of Sommerfeld's method in algebraic terms, and with the aid of Riemann's mapping theorem, leads to a readily generalized procedure. This reformulation has been applied to the analysis of diffraction in the text.

Sommerfeld used a distinctly different method to construct multiple-valued solutions to the equation of wave propagation. It is lengthy and is only sketched below. It does not appear to have been generalized: Ref. [5] asserts that the method used "requires remarkable insight and a very through knowledge of the different types of solutions to the wave equation, and the solution of the diffraction problem must be regarded as a triumph of ingenuity and experiment."

In order to analyze the diffraction of a plane wave by a semi-infinite plane, Sommerfeld first constructs a spherical harmonic (a solution to Laplace's equation in ordinary single-leaved space) of general form:

$$\frac{(-1)^{m=1}}{m!}\frac{\partial^m}{\partial z^m}\left[\frac{1}{D_o}F\left(\frac{x+iy}{z+D_o}\right)\right], \tag{B.6}$$

where x, y, and z are coordinates of the point of observation in a rectangular Cartesian coordinate system, $D_o^2 = x^2 + y^2 + z^2$, and m is a positive integer. Equation (B.6) represents a spherical harmonic of order $-m-1$. The function F is analytic but otherwise arbitrary, allowing enough latitude to define a multiple-valued function in three-dimensional space.

In order to replace repeated differentiations by an algebraic expression, Eq. (B.6) can be written as a Cauchy integral:

$$\frac{1}{2\pi i}\int_C \frac{dw}{D_1}F\left(\frac{x+iy}{w+D_1}\right)\frac{1}{(z-w)^{m+1}}, \tag{B.7}$$

where $D_1^2 = x^2 + y^2 + w^2$ and C is a closed contour in the complex w-plane containing $w=z$ in its interior. Evidently, the possibility of choosing the contour C in a convenient manner lends further flexibility to the method of construction.

Replacing z with w, a complex number, implies that D_1 now has a branch point $w_b = \pm i\sqrt{x^2+y^2}$ and this interferes with the free choice of desired multiple valued properties of the function F. To avoid this difficulty, a term is subtracted from the integrand in Eq. (B.7) to obtain

$$V_{-m-1} = \frac{1}{2\pi i}\int_C \frac{dw}{D_1}\frac{1}{(z-w)^{m+1}}\left[F\left(\frac{x+iy}{w+D_1}\right) - F\left(\frac{x+iy}{w-D_1}\right)\right]. \tag{B.8}$$

Next, with the substitution $w = i\left(+\sqrt{x^2+y^2}\right)\cos u$, the integral in Eq. (B.7) becomes

$$\frac{1}{2\pi}\int_C \left[\frac{du}{\left(z-\sqrt{z^2-1}\cos u\right)^{m+1}} \right] \left\{ F\left[\exp i\left(\phi-u-\frac{\pi}{2}\right)\right] - F\left[\exp i\left(\phi-u-\frac{\pi}{2}\right)\right] \right\},$$

$$(B.9)$$

where the contour C_1 encloses a point $u=u_o$ such that $z=i|\rho|\cos u_o$ and $\rho=x+iy$, $\exp i\phi=\rho/|\rho|$.

Now transformations $x=(k/m)x_T$ and $y_l=(k/m)y_T$ are made, assuming that $x^2+y^2+z^2=1$ and k is a positive constant. Upon increasing m without bound, one finds that the function V_{-m-1} becomes[1,3]

$$\frac{1}{2\pi}\int du \left\{ F\left[\exp i\left(\phi-u-\frac{\pi}{2}\right)\right] - F\left[\exp i\left(\phi+u-\frac{\pi}{2}\right)\right] \right\} \exp ikr \, \cos u,$$

$$(B.10)$$

where the subscript T has been dropped and $r=+\sqrt{x^2+y^2}$. With a suitable choice of contour, it can be shown that the function in Eq. (B.10) is a solution to the wave propagation equation.[1,3]

With the choice of

$$F(\rho)=\frac{1/2}{1-(\rho/\rho')^{1/2}},$$

$$(B.11)$$

where $\rho=x+iy$ and $\rho'=\exp i(\phi'-\pi/2)$ with $0<\phi'<\pi/2$, one obtains the desired solution representing a plane wave incident on a semi-infinite plane.

For a more complete treatment, the reader is referred to Refs. [1] and [3]. At this point, however, it can be seen that the method proposed in the text is more direct and straightforward than the method of analysis of diffraction used by Sommerfeld.

References

1. A. Sommerfeld, "Mathematische Theorie der Diffraction," *Math. Ann.*, **47**, pp. 317–374 (1896).

2. A. Sommerfeld, "Über verzweigte Potentiale im Raum," *Lond. Math. Soc. Proc.*, **28**, pp. 395–429 (1897).

3. F. Alzofon, *Multiple-Valued Functions in Three-Dimensional Space and Sommerfeld's Method*, Lockheed Electronics Co., Houston, TX (1970).

4. S. Neustadter, "Multiple-valued harmonic functions with circle as branch curve," *Univ. Calif. Publ. in Math.*, University California Press, New Series, **1**, pp. 281–340 (1951).

5. H. Bateman, *Partial Differential Equations of Mathematical Physics*, Dover Publications, New York (1944).

Appendix C
Analytic Functions of a
Complex Variable

It is the purpose of this appendix to indicate, very briefly, the analytic basis for the theory used in this book. No mathematical proofs are offered; these are to be found in the references cited.

C.1 Complex Numbers

A complex number z is the sum of a real number x and an imaginary number iy, where $i = \sqrt{-1}$ and y is a real number. The complex conjugate $z*$ is equal to $x - iy$. Two complex numbers $a + ib$ and $c + id$ are equal if and only if $a = c$ and $b = d$. If $z = x + iy$, then $|z| = +\sqrt{x^2 + y^2}$.

A complex function $f(z)$ is the sum of the two functions $u(x,y)$ and $iv(x,y)$—that is $f(z) = u(x,y) + iv(x,y)$—where both u and v are real functions.

C.2 Differential Properties

The derivative $(df/dz) = f'(z)$ of a complex function $f(z)$, at a fixed point $z_o = x_o + iy_o$, is defined as the unique limit (if it exists)

$$\lim_{\Delta z \to 0} \frac{f(z_o + \Delta z) - f(z)}{\Delta z}, \tag{C.1}$$

as Δz tends to zero from any direction. That is, the ratio tends to the same limit, for example, if $\Delta x = 0$ and Δy tends to zero, and if $\Delta y = 0$ and Δx tends to zero as well. Symbolically,

$$-i\frac{\partial u}{\partial y} + \frac{\partial v}{\partial y} = \frac{\partial u}{\partial x} + i\frac{\partial v}{\partial x}. \tag{C.2}$$

It then follows that

$$\frac{\partial u}{\partial x} = \frac{\partial v}{\partial y} \tag{C.3}$$

and

$$\frac{\partial u}{\partial y} = -\frac{\partial v}{\partial x}. \tag{C.4}$$

These equations are called the Cauchy-Riemann equations, and the conjugate functions u and v satisfy these equations. A similar pair of conjugate functions, η and θ, are used extensively in the text.

If $f(z)$ has a derivative in a region of the z-plane, it is termed analytic in that region. The function $f'(z)$ is finite and single valued in the region.

A remarkable consequence of analyticity is that an analytic function possesses derivatives of all orders in the region, and it can be developed in a Taylor series that is valid in the neighborhood of each point z_o in the region:

$$f(z) = f(z_o) + (z - z_o)f'(z_o) + \left[(z - z_o)^2 / 2!\right] f''(z_o) + \dots \tag{C.5}$$

C.3 Integral Properties of Analytic Functions

An analytic function can be written as a Cauchy integral in the region of analyticity as

$$f(z) = \frac{1}{2\pi i} \int_C dt \frac{f(t)}{t - z}, \tag{C.6}$$

where the integral is taken over a closed contour C in the region, described in a counterclockwise sense and containing the point z in its interior. It can then be shown that

$$f'(z) = \frac{1}{2\pi i} \int_C dt \frac{f(t)}{(t - z)^2} \tag{C.7}$$

$$\vdots \qquad \vdots$$

$$f^{(n)}(z) = \frac{n!}{2\pi i} \int_C dt \frac{f(t)}{(t - z)^{n+1}}; \tag{C.8}$$

i.e., in effect the latter equations assert that it is possible to differentiate "under the integral sign."

C.4 Singularities

The analytic property may fail in a variety of ways. Those of interest here are

(1) The function $f(z)$ may have poles, i.e., $f(z)$ can be expanded in a series of the form

$$f(z)=a_o+\frac{a_{-1}}{z-z_o}+\cdots\frac{a_{-m}}{(z-zo)^m}, \tag{C.9}$$

where m is a positive integer. In such a case, $f(z)$ is said to have a pole of order m at the point z_o. The function $f(z)$ may be analytic in a region excluding the point z_o.

(2) If the expansion of Eq. (C.9) is an infinite series, z_o is called an essential singularity.

(3) The function $f(z)$ may be multiple valued at the point z_o, i.e., a single circuit of z_o by a variable point does not return $f(z)$ to its original value. An example of such a function is $(z-z_o)^{1/2}=|z-z_o|^{1/2}\exp i\psi/2$, where $z-z_o=|z-z_o|\exp i\psi$. The singularity can be removed by introducing a barrier in the z-plane so that $f(z)$ becomes single-valued and the theorems about analytic functions can be applied. The point z_o is called a branch point.

C.5 Contour Integration

The concept of integration contour deformation is essential in the construction of the Sommerfeld solutions. The concept is based on the theorem that for two contours of integration C_1 and C_2 between the same two fixed points, in a region in which $f(z)$ is analytic, the integrals of $f(z)$ over the two contours are equal in value.

For a closed contour containing a pole in its interior, i.e., the function $f(z)$ can be expanded in a series:

$$f(z)=\left[\frac{a_{-m}}{(z-z_o)^m}\right]+\left[\frac{a_{-m+1}}{(z-z_o)^{m-1}}\right]$$

$$+\cdots+\left[\frac{a_{-1}}{z-z_o}\right]+a_o+a_1(z-z_o)+\cdots, \tag{C.10}$$

in the neighborhood of the pole z_0; then the integral about a closed curve containing z_0 in its interior, described in a counterclockwise sense, is equal to $2\pi i a_{-1}$. The constant a_{-1} is called the residue of $f(z)$.

If there are several such points, z_0, z_1..., then the contour integral is equal to the sum of the residues at the latter points, multiplied by $2\pi i$.

C.6 Analytic Continuation

In the limited sense used in the text, the term "analytic continuation" refers to the following definition:

For an analytic function $f(x)$ defined for x on an interval of real numbers, x is replaced by a complex number z defined on a region of the z-plane that includes the interval on which x is defined. The resulting function $f(z)$, defined on the larger region of the z-plane, is called the analytic continuation of $f(x)$. Evidently, $f(z) = f(x)$ if the values of z are restricted to the interval on which x is defined.

References

1. K. Knopp, *Theory of Functions, Part One*, translation by F. Bagemihl, Dover Publications, New York (1945).
2. N. McLachlan, *Complex Variable Theory and Transform Calculus, Second Edition*, Cambridge University Press, New York (1955).

Appendix D
Uniform Convergence

The Sommerfeld construction of a multiple-valued function, both for solutions of Laplace's equation and the equation of wave propagation, leads to a contour integral with part of the contour at infinity. Subsequently, it is necessary to verify that the function derived satisfies the latter equations; this requires what is often called "differentiation under the integral sign." To justify such a process, it is not enough that the integral converge as part of the contour recedes to infinity; to ensure that differentiation can take place, it is sufficient that the integral converge *uniformly*.

D.1 Definition of Uniform Convergence

For simplicity, consider a real function of two variables $f(x,y)$, where both x and y are real $(a \leq x \leq b)$ and assume that the function $f(x)$ is defined by the convergent integral

$$f(x) = \int_0^\infty dy \, f(x,y). \tag{D.1}$$

The integral is said to converge uniformly if the remainder of the integral

$$\left| \int_c^\infty dy \, f(x,y) \right| \tag{D.2}$$

can be made arbitrarily small if the lower limit c can be chosen large enough: a choice valid for all values of x in the interval $a \leq x \leq b$.[1]

An important criterion for uniform convergence is that $(a \leq x \leq b, p > 1, \text{ and } M > 0)$

$$|f(x,y)| < M \, y^{-p} \tag{D.3}$$

for large enough y.[1] The latter criterion can be replaced by $(q < 0)$

$$|f(x,y)| < M \exp - qy, \tag{D.4}$$

since, for large enough y, $\exp(-qy) < y^{-p}$ [note that $\lim\limits_{y\to\infty} y^p \exp(-qy) = 0$]. The latter criterion corresponds to the integrands encountered in the text, which, for large enough $|\eta''|$, vary essentially as $\exp{-q|\eta''|}$.

References

1. R. Courant, *Differential and Integral Calculus, Vol. II*, translation by E. McShane, Interscience Publishers, New York (1936).
2. H. Carslaw, *Introduction to the Theory of Fourier's Series and Integrals, Third Edition*, Macmillan and Co., London (1930).

Frederick E. Alzofon is a consultant in electro-optics, now retired from a 30-year career as an electro-optics engineer in the aerospace industry. He received a B.A. degree from the University of California at Los Angeles, and M.A. and Ph.D. degrees from the University of California at Berkeley. His research interests have included the theory and practice of thermography, of which he is one of the founders. A lasting interest, derived from thermography, has been the development of optical analogues in the analysis of heat and viscous fluid flow, as well as energy propagation in dispersive mediums. He has published a book, papers, and reports on the following subjects: Sommerfeld's method for potentials in three-dimensional space; experimental investigations into the properties of gases with use of a shock tube; analysis of electron capture by protons; analysis of high-energy collisions of neutrons and protons; the theory of relativity; a unified field theory; engineering applications of gravity control; and several papers on exchange blood transfusions.